DECODE TO ENCODE

ABOUT THE AUTHOR

Avinash Ramachandran has been working in video compression for over 15 years as a serial startup technologist, innovator, and speaker. His work on video coding has contributed to several patents in motion estimation, motion compensation and bitrate control algorithms. He is currently developing next-generation algorithms and products with H.265, VP9 and AV1 codecs at NGCodec Inc. A senior member of IEEE, he completed his Masters in Digital Signal Processing from the Indian Institute of Technology Madras in India and holds an MBA from the Richard Ivey School of Business in Canada.

AVINASH RAMACHANDRAN

DECODE TO ENCODE

MASTER COMPLEX CONCEPTS FASTER, BRIDGE GAPS
AND BE THE EXPERT IN VIDEO CODING

DECODE TO ENCODE

*To my wife **Veena**, who's always lovingly accepted me, supported me and stood by me.*

LIST OF FIGURES

LIST OF TABLES

ACRONYMS

This list includes the acronyms used in the book, listed alphabetically.

AV1	AOMedia Video1
AVC	Advanced Video Coding
AOM	Alliance for Open Media
AR	Augmented Reality
BO	Band Offset filter
CABAC	Context Adaptive Binary Adaptive Coding
CAVLC	Context Adaptive Variable Length Coding
CB	Coding Block
CBR	Constant Bitrate
CPB	Coded Picture Buffer
CRF	Constant Rate Factor
CTB	Coding Tree Block
CTU	Coding Tree Unit
DCT	Discrete Cosine Transform
DPB	Decoded Picture Buffer
DST	Discrete Sine Transform
EG	Exp-Golomb code
EO	Edge Offset filter
fps	frames per second

GOP	Group of Pictures
HD	High Definition
HDR	High Dynamic Range
HEVC	High Efficiency Video Coding
HRD	Hypothetical Reference Decoder
HVS	Human Visual System
IDCT	Inverse Discrete Cosine Transform
IDR	Instantaneous Decoder Refresh
ISO	International Organization for Standardization
JCT	Joint Collaborative Team (ISO and ITU)
JVT	Joint Video Team
KLT	Karhunen-Loeve Transform
MAD	Mean Absolute Difference
MC	Motion Compensation
MPEG	Moving Picture Experts Group
ME	Motion Estimation
ML	Machine Learning
MR	Mixed Reality
MV	Motion Vector
MVD	Motion Vector Difference
OTT	Over the Top

PSNR	Peak Signal to Noise Ratio
QP	Quantization Parameter
RC	Rate Control
RDO	Rate-Distortion Optimization
RGB	Red Green Blue (Color format)
SAD	Sum of Absolute Differences
SAO	Sample Adaptive Offset filter
SATD	Sum of Absolute Transformed Differences
SD	Standard Definition
SDR	Standard Dynamic Range
SEI	Supplemental Enhancement Information
SSIM	Structural SIMilarity
TR	Truncated Rice
UHD	Ultra-High Definition
VBR	Variable Bitrate
VCEG	Visual Coding Experts Group
VMAF	Video Multimethod Assessment Fusion
VOD	Video on Demand
VQ	Video Quality
VR	Virtual Reality
YCbCr	Color Format with Luma (Y) and two Chroma (Cb and Cr)

YUV   Color Format with Luma (Y) and two Chroma (U and V)

PREFACE

Video coding is complex. YouTube and Netflix use it to deliver great video even in extreme network transmission conditions, but have you ever wondered how they optimize video for low bandwidths? Do technical terms like rate distortion optimization, predictive coding or adaptive quantization overwhelm you? Have you tried to understand video compression but felt confused and frustrated about where to start?

This is a comprehensive book that can break through any barriers you may have to understand such technological matters. The chapters and sections consolidate fundamental video coding concepts in an easy-to-assimilate structure. *Decode to Encode* is the only book that has been designed to answer the hows and whys of elements of the H.264, H.265, and VP9 video standards. It explains the common coding tools in these three successful standards in clear language, providing examples and illustrations as much as possible.

The book neither pertains to any specific standard nor attempts to show if one standard is better than any other. It provides video engineers and students with the understanding of compression fundamentals underlying all major standards that they need to help solve problems, conduct research and serve their customers better.

I have been a lifetime student of video coding and an active contributor to the development of encoding algorithms and software based on the MPEG2, H.264, H.265 and VP9 standards for encoding and decoding. When researching for this book, I drew on years of personal experience as a video codec engineer and product manager. I've also talked with numerous experts in the industry and drawn on information in books and online material on video compression topics. These are compiled in a list of resources at the end of this book.

Knowledge is power, and time is money. Video professionals, students, and others can use this book to quickly build a solid foundation and become experts in next-generation video technologies. Managers and leaders in the industry can use it to build expert teams and significantly boost productivity and creativity. Video technology has advanced rapidly, and compression has improved by over 500% across several generations of standards. Still, the key concepts that have formed the backbone and framework of video and image compression remain the same after four decades of progress. The book you hold in your hands is the one I wish I had when I was a beginner in video coding back in the early 2000s.

Caleb Farrand, a software engineer from San Francisco has said "With this book, much of the vocabulary I'd hear around the office started to make sense and I got a better understanding of what encoders do and how they are designed"

Why be the person who gets left behind because you are just drowning in everyday work and don't have the time? Instead, transform your career roadmap today by building your foundations to better understand the changing video technology landscape and be a part of it. Become the person others in the industry look up to for expertise.

I promise that once you understand the concepts dealt with in this book, you will feel significantly more confident and like an expert as you walk into the next meeting with peers or customers. AND I promise you will be inspired more than ever to explore advanced topics in video compression and take up initiatives that just seem too daunting and far-fetched right now. If you desperately need momentum to leapfrog to your career goals, this book can help.

Everything you need to understand video coding concepts is available to you right here. The compression topics have been carefully organized along the common thread that ties together all the major coding standards like H.264, H.265, and VP9. This makes the information easy to absorb, even in the middle of your busy schedule, travels, and agile work environment. As you keep reading, you will find that each chapter will give you new insights into the next steps in the compression pipeline and equip you with the tools you need to understand how newer video coding technologies are built. This will, in turn, help you to understand how video can be optimized to meet the requirements of emerging experiential technologies like 360° video, virtual reality, augmented reality and mixed reality. The video technology space is more exciting than ever and if you are working in this space, the possibilities to grow are endless!

Avinash Ramachandran
November 2018

ACKNOWLEDGEMENTS

I would like to acknowledge everyone whose help I have benefitted from in many ways in bringing shape to this book. The examples presented in the book use the raw video *akiyo* (CIF), *in_to_tree* (courtesy SVT Sweden), *Stockholm* and *DOTA2* sequences hosted on the Xiph.org Video Test Media [derf's collection] website. The work and efforts of experts in standards organizations, companies and associations like JCT-VC, IEEE, Fraunhofer HHI, Google and universities driving state-of-the-art research in Video Compression technologies have contributed significantly to this book. I would like to thank authors and contributors of several excellent books on the subject, online technical blog articles and research papers whose material I consulted while writing this book. An extensive list of these sources is also included in the resources section of the book. Special thanks for review, discussions, and support are also due to Rakesh Patel, Oliver Gunasekara, Yueshi Shen, Tarek Amara, Ismaeil Ismaeil, Akrum Elkhazin, Harry Ramachandran and Edward Hong. Thanks to Susan Duncan for all the editorial assistance. This book has been possible only because of the unstinting support of my wonderful family: my mother Vasanthi, my wife Veena and our children Agastya and Gayatri.

ORGANIZATION OF THE BOOK

This book is organized into three parts. Part I introduces the reader to digital video in general and lays the groundwork for a foray into video compression. It provides details of basic concepts relevant to digital video as well as insights into how video is compressed and the characteristics of video that we take advantage of in order to compress it. It also covers how exactly these characteristics are exploited progressively to achieve the significant level of compression that we have today. Part I concludes by providing a brief history of the evolution of video codecs and summarizes the important video compression standards and their constituent coding tools.

Building on this foundation, Part II focuses on all the key compression technologies. It starts by covering in detail the block-based architecture of a video encoder and decoder that is employed in all video coding standards, including H.264, H.265, VP9, and AV1. Each chapter in Part II explains one core technique, or block, in the video encoding and decoding pipeline. These include Intra prediction, inter prediction, motion compensation, transform and quantization, loop filtering and rate control. I have generously illustrated these techniques with numerical and visual examples to help provide an intuitive understanding. Well-known, industry-recognized clips, including *in_to_tree*, *stockholm*, *akiyo*, and *DOTA2* have been used throughout the book for these illustrations. I have also attempted to provide explanations of not just the overall signal flow but also why things are done the way they are.

Equipped with all the essential technical nuts and bolts, you will then be ready to explore, in Part III, how all these nitty-gritties together make up an encoder, how to configure one and how to use it in different application scenarios. Part III presents different application scenarios and shows how encoders are tuned to achieve compression using the tools that were detailed in Part II. Specifically, the section explains in detail the various bit rate modes, quality metrics and availability, and performance testing of different codecs. Part III concludes with a chapter on upcoming developments in the video technology space, including content-specific, per-title optimized encoding, the application of machine learning tools in video compression, video coding tools in the next generation AV1 coding standard, and also compression for new experiential video platforms like 360 Video and VR.

I hope that the book is able to illustratively convey the entire video compression landscape and to inspire the reader toward further exploration, collaborations, and pioneering research in the exciting and rapidly-advancing field of video coding.

PART I

1 INTRODUCTION TO DIGITAL VIDEO

In this chapter, we explore how visual scenes are represented digitally. I will explain various specialized terms used in digital video. This is useful before we explore the realm of digital video compression. If you have a working knowledge of uncompressed digital video, you may briefly skim through this section or skip it entirely and proceed to the next chapter. Once you complete this chapter, you will better understand how terms like sampling, color spaces and bit depths apply to digital video.

Digital video is the digital representation of a continuous visual scene. In its simplest sense, a visual scene in motion can be represented as a series of still pictures. When these still pictures are consecutively displayed in rapid progression, the human eye interprets the pictures as a moving scene rather than perceiving the individual images. This is why, during the early days of filmmaking, it was called *movi*ng pictur*es*. This, over time, became condensed to *movies*.

$$\text{MOVIES} = \textit{Movi}\text{ng} + \text{Pictur}\textit{es}$$

To capture a visual scene and represent it digitally, the cameras therefore temporally sample the scene; that is, they derive still images from the scene at intervals over time. The method of capturing and displaying a complete picture at regular intervals of time results in what is referred in the industry as *progressive* video. Also, every temporal image is spatially sampled to get the individual *digital pixels*.

1.1   INTERLACED VIDEO

In the early days of television, *interlaced* video technology was used to represent video images that were projected on CRTs. In interlaced video, every line comprising a row of pixels is scanned to make up a picture. Each of these lines is called a *field* and alternate field lines are scanned and displayed in succession. Every odd field is scanned first followed by every even field. Each of these fields is displayed in half the time used to display a complete frame. Thus, a single frame of video is scanned and displayed as two half frames that are interwoven with one another. This is shown in Figure 1. As these fields are displayed at half the time it takes to display a frame, it happens very quickly, and we get the illusion of a full frame.

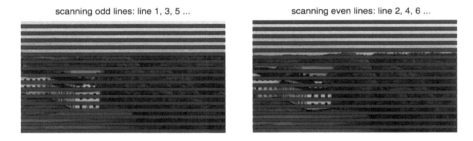

Figure 1: Scanning odd and even lines in interlaced video

Today, we have progressed from analog video to high definition digital video. However, legacy interlaced video formats still exist in linear video broadcasting. When interlaced content is used, the video is represented with an 'i' after the video resolution e.g. 1080i, 480i. We have briefly discussed interlaced video mechanism in this section to provide a good background. The remainder of the book will focus exclusively on progressive video technology.

1.2   SAMPLING

So, what is *sampling*? Sampling is the conversion of a continuous signal to a discrete signal. This can be done in space and time. In case of video, the visual scene sampled at a point in time produces a frame or picture of the digital video. Normally, scenes are sampled at 30 or 25 frames in one second, however, a 24 frames-per-second sampling rate is used for movie production. When the frames are rapidly played back at the rate at which they were sampled (frames per second or fps), they produce the motion picture effect. Each frame, in turn, is composed of three *components*, one of which is usually needed to represent monochrome pictures and the remaining two are included only for color images. These components are obtained by sampling the image spatially and together are called *pixels* or *pels*.

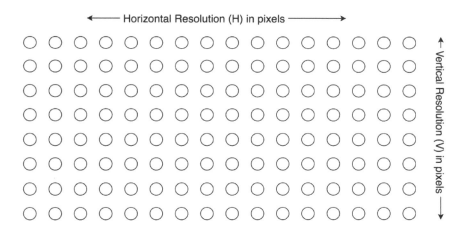

Figure 2: Horizontal and vertical resolutions of a picture.

Thus, every pixel in a video has one component (for monochrome) or three components (for color). The number of spatial pixels that make a video frame determines how accurately the source has been captured and represented. As shown in Figure 2, this is a 2-dimensional array of horizontal and vertical sample points in the visual scene and the total number is the parameter called *video resolution*. Mathematically, this can be expressed as:

Resolution = H (in pixels) x V (in pixels)

Thus, if a video has a resolution of 1920x1080, this means that it has 1920 horizontal pixel samples and 1080 vertical pixel rows. It should be noted that the resolution of the video usually refers to the first component of the video, namely, *luminance,* while the two color, or *chrominance,* components may be sampled at the same or lower sampling ratios. The resolution, combined with the frame capture rate expressed in frames per second, determines the captured digital image's degree of fidelity to the original visual scene. In turn, this also determines how much processing and bandwidth is needed to efficiently encode the video for transmission and storage.

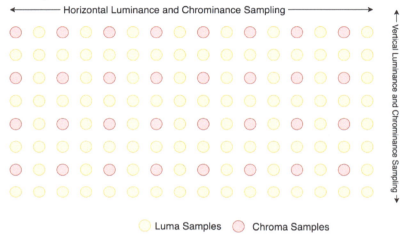

Figure 3: Luminance and chrominance spatial sampling in a picture.

Figure 3 shows how both the brightness component (called luminance) and color components (chrominance) are sampled in a typical picture. In this example, the chrominance is subsampled by a factor of 2, both horizontally and vertically, compared to the luminance. Figure 4 illustrates how the temporal sampling into pictures and the spatial sampling into pixels comprise the digital representation of an entire video sequence.

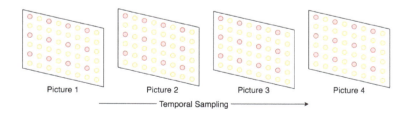

Figure 4: The spatial and temporal sampling of a visual scene.

When a video has to be compressed (by a process called *encoding*), the resolution can be changed, depending on various factors like bandwidth availability for storage or for transmission. To fit within these constraints, either the bitrate of the encoded video or the video resolution or a combination of these is adjusted to ensure the final encoded video can be delivered within the constraints of the overall system. Usually, the captured video source is downscaled to the required resolution and then encoded.

This is illustrated in Figure 5, which shows a frame from the *in_to_tree* video sequence [1]. The clip has been downscaled at different resolutions. In internet video streaming, the video is often encoded at multiple resolutions, each of which can serve requests from users who have different bitrate allocations. The following table outlines commonly used video resolutions and bandwidths for some applications.

Table 1: Common video resolutions and bandwidths across applications.

Resolution	Typical Bitrates	Applications
320x240 at 30fps	200 kbps – 500 kbps	Mobile video
720x480 at 30fps 720x576 at 25fps	500 kbps – 2 Mbps	Storage (DVD) & broadcast TV transmission
1280x720 at 30 and 25fps	1 Mbps – 3 Mbps	Video calling, internet video streaming
1920x1080 at 30 and 60fps 1280x720 at 60fps	4 Mbps – 8 Mbps	Internet video streaming, storage and broadcast transmission

Figure 6 shows a 16x16 block in the frame that, when zoomed in, clearly shows the varying color shades in the 16 x 16 square matrix of pixels. Every small square block in this zoomed image corresponds to a pixel that is composed of three components with unique values.

Figure 5: A frame of video downscaled to different resolutions [1].

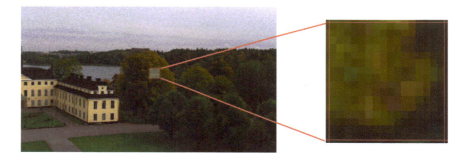

Figure 6: A 16x16 block of pixels in a frame that illustrates the 256 square pixels [1].

1.3   COLOR SPACES

Colors that make up visual scenes in the real world need to be converted to a digital format in order to represent the visual scene as a series of pictures. A *color model* is a mathematical model that converts any color to numerical values, so it can be processed digitally. Using the color model, a color can be represented as combinations of some base color components.

A *color space* is a specific implementation of a color model that maps colors from the real world to the color model's discrete values. Adding this mapping function provides a definite area or range of colors that is supported by the color space and this is called a *color gamut*. Two popular color spaces are used to represent video, namely, RGB and YUV. These are interchangeable using mathematical functions. That is, one representation can be derived from the other.

RGB: As the name implies, in this model, every color is represented as a combination of red (R), green (G) and blue (B) components. Each of these components is independent and combinations of these three components produce the various shades of the color space.

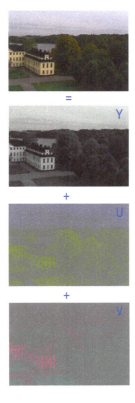

Figure 7: A color image and its three constituent components [1].

YCbCr: In this scheme which is often also referred to as YUV, the Y component refers to luma or intensity and Cb and Cr refer to chroma or color components. This scheme of representation of pixels is the most popular for video applications due to the following characteristics that help to reduce the bits needed for the representation:

- In this scheme, luma and chroma components are expressed completely independently of each other. This means, in the case of monochrome video, only one component (Y) is required to completely represent the video signal.

- The human visual system (HVS) is more sensitive to luma (Y) and less sensitive to chroma (UV). By emphasizing luma and selectively ignoring chroma components, significant reductions may be obtained without a huge impact on the viewer's experience of the video. This can be achieved by *subsampling* chroma relative to luma. Subsampling will be covered in more detail in Chapter 2.

In this book, we will be dealing with YUV video format exclusively for all video explanations unless otherwise stated. As illustrated in Figure 7, the Y component in the image corresponds to the intensity. This can be used to represent the monochrome version of the image or video whereas the U and V components together constitute the color components.

1.4   BIT DEPTH

The number of bits used to represent a pixel determines how accurately the visual information is captured from the source. It also determines the intensity variation and range of colors that can be expressed with the sample. That is, if only 1 bit were used to store a pixel value, it could have a value of either 0 or 1. As a consequence, only two colors could be expressed using this pixel: black or white. However, if 2 bits were used then every pixel could represent any of 4 (or 2^2) colors with values 0, 1, 2 and 3.

Video pixels are usually represented using 8 bits per sample. This allows for 256 (or 2^8) variations in color and intensity with values in the range of 0 to 255. However, the normal practice is to restrict the active luminance to a range of 16 (black) to 235 (white). Values 1–15 and 236–254 are reserved for *foot room* and *head room*, respectively, during studio production.

In professional video production systems, a video sequence is processed using 10 bits per sample. This allows 1024 gradations, such that much more subtle levels in intensity and color can be captured. 10-bit encoding is also becoming increasingly popular for UHD resolutions and HDR functionality to provide a richer visual experience in consumer video systems. This is also supported in modern video compression standards.

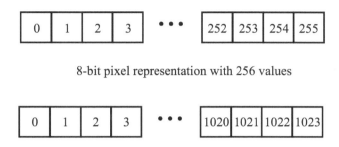

8-bit pixel representation with 256 values

10-bit pixel representation with 1024 values

Figure 8: 8-bit and 10-bit representation of pixel values.

1.5   HDR

Video technology continues to evolve from high definition (HD) resolutions to ultra HD (UHD) and beyond. The new technologies offer four or more times the resolution of HD. Given this evolution, it becomes important to better represent this high density of pixels and the associated colors in order to achieve the enhanced viewing experience that they make possible. Various methods have been explored to improve the representation of a digital video scene, including these two:

1. **Improved spatial and temporal digital sampling.** This includes techniques to provide faster frame rates and higher resolutions, as mentioned above.

2. **Better representation of the values of the pixels.** Several techniques are incorporated in HDR video to improve how various colors and shades are represented in the pixels.

While the former mostly deals with different ways of pixel sampling, the latter focuses on every individual pixel itself. This is an enhancement over the traditional *standard dynamic range* (SDR) video. *High dynamic range* video provides a very significant improvement in the video viewing experience by incorporating improvements in all aspects of pixel representations. In this section, we shall explore how this is done.

HDR, as the name indicates, provides improved dynamic range, meaning that it extends the complete range of luminance values, thereby providing richer detail in terms of the tone of dark and light shades. It doesn't stop there but also provides improvements in the representation of colors as well. The overall result is a far more natural rendering of a video scene.

It's important to note that HDR technology for video is quite different from the technology used in digital photography that also uses HDR terminology. In photography, different exposure values are used, and the captures are blended to expand the dynamic range of pixels by creating several local contrasts. However, every capture still uses the same 8-bit depth and 256 levels of brightness. HDR in video extends beyond just the dynamic range expansion to encompass the following [2]:

- high dynamic range with higher peak brightness and lower black levels, offering richer contrast;

- improved color space or *wide color gamut* (WCG), specifically, a new color standard called Rec. 2020 that replaces the earlier Rec. 709 used in SDR;

- improved *bit depth, either* 10-bit (distribution) or 12-bit (production) used instead of traditional 8-bits in SDR;

- improved *transfer function, for instance,* PQ, HLG etc. used instead of the earlier gamma function;

- improved *metadata,* in that HDR includes the addition of static (for the entire video sequence) and dynamic (scene or picture-specific) metadata which aid in enhanced rendering.

Table 2: Summary of enhancements in HDR video over earlier SDR.

Features	SDR	HDR
Dynamic range	Standard	Enhanced dynamic range with high peak brightness and lower black levels and greater contrast
Bit depth	8-bit	10-bit or 12-bit

Color space	REC.709	Rec.2020
Transfer function	Gamma based	Different new standards: PQ, HLG etc
Metadata	Not present	Static or dynamic

In the remainder of this chapter, we shall explore each of these enhancements briefly and consider how they apply to video encoding technology.

1.5.1   COLOR SPACE

In the early 1990s, the HDTV standard was established with the color gamut defined by a standard called Rec. ITU-R BT. 709 (popularly known as Rec. 709). This was enhanced for UHD in 2012 under Rec. ITU-R BT. 2020 (Rec. 2020) to provide a far larger color space that supports a greater variety of shades.

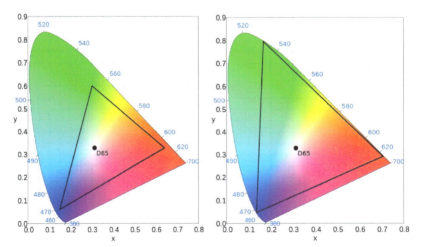

Figure 9: Color ranges in Rec. 709 (left) versus Rec. 2020 (right).
Source: https://commons.wikimedia.org/wiki/File:CIExy1931.svg [3]

Figure 9 compares the color spaces available under Rec. 709 versus Rec. 2020. The black triangle drawn within each represents the coverage of color shades under that standard. Clearly, Rec. 2020 supports a more expanded color gamut with very many more shades. This is what is implemented in HDR.

1.5.2   BIT DEPTH

Bit depth, as was explained in the previous section, is the number of bits used to represent every pixel. This, in turn, determines the total number of colors that can be represented. Traditional SDR Video uses 8-bit depth, meaning that every pixel can have 256 different values each for red, green and blue (or correspondingly, for Y, U, and V). This results in a total of 256 x 256 x 256, or about 16 million colors per pixel.

HDR supports 10-bit color for video distribution, meaning that every pixel now can represent up to 1026 values per color component or over a billion colors. This massive increase packs in a far more extensive range of shades. This results in smoother color transitions within the same color groups with few artifacts. Thus, increasing the number of pixels even by 2 bits is extremely useful in improving the visual experience. It does bring with it, however, increased memory requirements and computational expense during any internal process, such as encoding.

1.5.3   TRANSFER FUNCTION

Electronic image devices and video capture and display devices have the need to convert electronic signals to digitally represented pixel values and vice versa. These devices have nonlinear, light intensity-to-signal or signal-to-intensity characteristics.

As an example, the voltage from a camera has a nonlinear relationship to the intensity (power) of light in the scene. Traditionally, as specified in the Rec. 709 color standard, the nonlinear mapping is accomplished using a power function called the *gamma transfer function*. This is expressed in the generic equation

$$output = input^{gamma}$$

where *gamma*, the exponent of the power function, completely describes this transfer function. The gamma curve, which simply modeled how older electronic display devices (cathode ray tubes, or, CRTs) responded to voltage, is no longer completely relevant for modern displays. It has been improved upon in newer functions like PQ (perceptual quantization) and HLG (hybrid log gamma) in HDR.

1.5.4   METADATA

Video signals are compressed before transmission using an encoder at the source. The compressed signals are then decoded at the receiving end. When the decoder receives the encoded stream and produces raw YUV pixels, these pixels will need to be displayed on a computer screen or display. This means the YUV values will need to be converted to the correct color space. This includes a color model and the associated transfer function. These define how the pixel values get converted into light photons on the display panels. Modern codecs have provisions for signaling the color space in the bitstream header using *supplemental enhancement information* (SEI) messages.

However, what happens when the display device doesn't support the color space with which the source is produced? In this case, it's important to use the source color space characteristics and convert color spaces at the decoder end to the display's supported format. This is crucial to ensure that colors aren't displayed incorrectly when displays are incompatible with the source color space.

The HDR enhancements support a mechanism to interpret the characteristics of encoded pictures and use this information as part of the decoding process, incorporating a greater variety of transfer functions. As explained earlier, if content produced using a specific transfer function at the source goes through various transformations in the video processing and transmission workflow and then gets mapped using another transfer function by the display device, the content ends up perceptibly degraded. HDR standards provide enhanced *metadata mechanisms* to convey the transfer function details from the source to the decoding and display devices.

1.5.5   HDR LANDSCAPE

The HDR landscape does not have one unified format. Instead, it has a few options that have been developed and deployed by different organizations. The following are the five main HDR formats. [2]

-   Advanced HDR (developed by Technicolor and Philips)

- HDR 10+ (SMPTE ST-2094-40) with dynamic metadata
- HDR 10 with dynamic metadata (SMPTE 2094-X)
- Dolby Vision
- Hybrid Log-Gamma (HLG) with no metadata

Table 3 compares and summarizes the features available in these HDR formats.

Table 3: Comparison and summary of features available in HDR formats.

Format	Metadata	Details
Advanced HDR	Dynamic	Technicolor and Philips
HDR 10+	Dynamic	Samsung and Panasonic
HDR 10	Dynamic	SMPTE 2094-10: Dolby SMPTE 2094-20: Philips SMPTE 2094-30: Technicolor SMPTE 2094-40: Samsung
Dolby Vision	Dynamic	Dolby Labs
Hybrid Log-Gamma	Not present	BBC and NHK

Different standards use different mechanisms for metadata transmission. Some standards just add metadata during distribution. Earlier standards also used static metadata in which fixed metadata information is generated during video creation and propagated through the video workflow for the entire stream. Other mechanisms use dynamic metadata which can be varied on a per-frame, per-scene basis.

For widespread HDR adoption, it's imperative that devices consistently carry the signal and associated information right from the source to display as part of the end-to-end system. HDR standardization efforts can go a long way toward ensuring HDR adoption at scale. Other important activities will include embedding HDR information like HEVC SEI in the encoding process and HDR support in display standards like HDMI.

1.6   SUMMARY

- Digital video is the digital representation of a continuous visual scene that is obtained by sampling in time to produce frames, which in turn is spatially sampled to obtain pixels.
- Colors in the real world are converted to pixel values using color spaces.
- The number of bits used to represent a pixel determines how accurately the visual information is captured from the source and is called bit depth which is often 8-bit or 10-bit for video.
- HDR technology improves the visual experience by enhancing pixel representation. It incorporates advanced dynamic range, higher bit depth, advanced color space, and transfer functions.

1.7   NOTES

1.   in_to_tree. xiph.org. Xiph.org Video Test Media [derf's collection]. https://media.xiph.org/video/derf/. Accessed September 21, 2018.
2.   *High Dynamic Range Video: Implementing Support for HDR Using Software-Based Video Solutions*. AWS Elemental Technologies. https://goo.gl/7SMNu3. Published 2017. Accessed September 21, 2018.
3.   I, Sakamura. File:CIExy1931.svg, CIE 1931 color space. Wikimedia Commons. https://commons.wikimedia.org/wiki/File:CIExy1931.svg. Published July 13, 2007. Updated June 9, 2011. Accessed September 20, 2018.

2 VIDEO COMPRESSION

As explained earlier, digital video is the representation of a visual scene using a series of still pictures. In the previous chapter, we saw how digital video is represented and explained concepts like sampling and color spaces. In this chapter, we explore why video needs to be compressed and characteristics in the video signal that are exploited to achieve this. In a nutshell, video compression primarily focuses on how to take the contiguous still pictures, identify and remove redundancies in them and minimize the information needed to represent the video sequence.

2.1 LANDSCAPE

Video compression is essential to store and transmit video signals. Typically, video originates from a variety of sources like live sports and news events, movies, live video conferencing and calling, video games and emerging applications like augmented reality/virtual reality (AR/VR). While some applications like live events and video calling demand real-time video compression and transmission, others, like movie libraries, storage, and on-demand streaming are non-real time applications. Each of these applications imposes different constraints on the encoding and decoding process, resulting in differences in compression parameters. A live sports event broadcast requires high quality, real-time encoding with very low encoding and transmission latency, whereas encoding for a video-on-demand service like Netflix or Hulu is non-real time and focuses on highest quality and visual experience.

To this effect, every video compression standard provides a variety of toolsets that can be enabled, disabled and tuned to suit specific requirements. All modern video compression standards, including MPEG2, H.264, H.265, and VP9, define only the toolsets and the decoding process. This is done to ensure interoperability across a variety of devices. Every decoder implementation must decode a compliant bitstream to provide an output identical to the outputs of other implementations operating on the same input. Encoder implementations, on the other hand, are free to choose coding tools available in the standard and tune them as part of their design, as long as they produce an output video that is standard-compliant. Encoder implementations can design and incorporate different preprocessing techniques, coding tool selection, and tuning algorithms as part of their design. This may result in dramatic differences in video quality from one encoder to another.

2.2   WHY IS VIDEO COMPRESSION NEEDED?

Video representation in its raw form takes a lot of bits. As an illustration, for an uncompressed, raw, UHD resolution video (3840x2160 pixels at 60fps) with 10 bits/pixel for 3 color components, the bandwidth needed would be: 3840 x 2160 x 60 x 10 x 3 = 14.92 gigabits per second (Gbps). It's not practical to transmit such data using today's internet bandwidth without any processing as bandwidth is at most a few tens or hundreds of megabits per second (Mbps).

As an example, if we have to transmit the UHD video with the bandwidth of 14.92 Gbps over a 15 Mbps link, it would need to be compressed by a factor of 1000. Also, a 5-minute video at this resolution would need 559 GB of storage space if stored in its raw format. Imagine attempting to download such video onto your phone or tablet devices. These have storage available only in the range of from 16GB to 300GB.

Clearly, storing and transmitting uncompressed video present huge practical challenges. It is almost impossible. It is estimated that 70% of the internet traffic is video. [1] This percentage is rapidly increasing. There are also continuous technological advances, permitting increases in video resolutions to 4K and 8K. These are, respectively, four times and eight times larger than 1080p.

Other technological improvements include increased frame rates to 90fps and beyond for emerging immersive applications and HDR implementations. This is 25-50% larger than SDR and provides an enhanced user experience. These continue to push the limits of today's video experiences and compression is the crucial binding glue that enables these technologies to be deployed at scale.

Advanced video compression tools that can optimize video are thus essential for keeping up with the rapid improvements in video infrastructure and for reliable video transmission to provide next-generation improvements to consumer video experiences.

2.3   HOW IS VIDEO COMPRESSED?

Video signals comprise pixels that contain nearly the same values from one pixel to the next. In other words, there is significant redundancy in representing pixel values in video. Compression techniques work to identify and carefully remove these redundancies, by carefully considering the human visual system model. Broadly, these redundancies can be classified into 2 types:

1.   Statistical redundancies
2.   Psycho-visual redundancies

Statistical redundancy refers to the inherent redundancy that exists in the distribution of pixel values within the video sequence. This can be due to the nature of the visual scene itself. For example, if a scene is static, then there is far more redundant information across pixels than is the case with highly textured content. Statistical redundancy can also manifest in the way that these pixels are finally encoded in the bitstream. Statistical redundancy thus can be classified into 3 types:

1.   Spatial pixel redundancy
2.   Temporal pixel redundancy
3.   Coding redundancy

2.3.1   SPATIAL PIXEL REDUNDANCY

In a video scene, the pixels that are close together within the same picture (or frame) are significantly similar. Figure 10, below, shows a frame from the *akiyo* video sequence.

We see in Figure 10 that the pixels comprising the newsreader's dress, those of her face, and those of the background are very similar.

Statistically, the pixels that are close together are said to exhibit strong correlation.

We can exploit this strong correlation and represent the video with only a differential from a base value, as this requires far fewer bits to encode the pixels. This is the core concept of how differential coding is applied in video compression. All modern video coding standards remove these spatial redundancies to transmit only the minimal bits needed to make up the residual video.

Figure 10: Illustration of spatial correlation in pixels in one frame of the akiyo sequence [2].

For example, in the group of 4x4 pixels in Figure 11, 8 bits are needed to represent values from 0 through 255. Every pixel will need 8 bits for its representation.

240	242	243	243		0	2	3	3
240	241	243	243		0	1	3	3
240	242	243	243		0	2	3	3
240	241	243	243		0	1	3	3

Figure 11: Illustration of spatial correlation in pixels.

Thus, the 4x4 block can be completely represented using 128 bits. We, however, notice that the pixel values in this block vary only by very small amounts. If we were to represent the same signal as a differential from a base value (say 240), then every pixel in the block can be represented using 8 bits for the base value and 2 bits (values from 0 through 3) for each of the remaining 15 values, resulting in a total of only 40 bits required for representation. The base value chosen is called a *predictor* and the effective differential values from this predictor are called *residuals*. This mechanism of using differential coding to remove spatial pixel redundancies within the same picture is called *intra picture coding*.

2.3.2   EXPLOITING TEMPORAL PIXEL CORRELATION

Figure 12 shows three consecutive pictures that are part of the *akiyo* news reader video sequence. In addition to spatial similarities, the successive pictures in this digital video sequence are strikingly similar and static for the most part, except for some noticeable differences around the eyes and mouth of the news reader. This can vary with content and there can be a variety of objects in the scene and movement of different objects in different directions across pictures. However, the core idea is the presence of strong similarities among the pictures, also known as *temporal correlation* across successive pictures. This is illustrated in Figure 12.

This strong correlation of pixels across nearby pictures can be exploited to encode successive video frames using only information that's new to the frame and removing redundant data that has already been coded in previous pictures. This is done by coding only differential values from pixels in the previous frames. The mechanism of using differential coding to remove temporal pixel redundancies is called *inter picture coding*.

Akiyo Sequence frame 10   Akiyo Sequence frame 11   Akiyo Sequence frame 12

Figure 12: Illustration of temporal correlation in successive frames of the akiyo sequence [2].

2.3.3   ENTROPY CODING

In the pixel matrix in Figure 11, above, all residual values need to be transmitted correctly in order to be able to deduce the original pixel values from them. If we were to assign 2 bits to represent each of the four residual values, we would need 2 bits x 16 values = 32 bits to transmit the matrix of residuals. However, upon closer observation, we find that some residual values occur more frequently than others in the matrix. For example, the residual value 3 occurs 8 times. Thus, instead of an equal assignment of bits for each of the 4 values, if we were to assign fewer bits to the most frequently occurring values and more bits to rarely occurring values, this could result in further reduction in the total number of bits needed.

This principle of utilizing statistical redundancies to achieve compression during bit allocation is called *entropy encoding*. This technique is completely *lossless*, meaning the original pixel data can be accurately constructed from the coded values. For example, if we use the matrix shown in Table 4 to assign bits to each of the four values, we can represent all the residual values using a total of 4 x 2 + 2 x 3 + 2 x 3 + 8 x 1 = 28 bits. It should be noted that the statistical redundancy elimination methods discussed above, in addition to being extremely helpful in compression, are all also completely lossless. In the following section, we will discuss a few methods which exploit psycho-visual redundancies which will result in lossy coding. These methods provide the bulk of compression gains achievable with modern video codecs.

Table 4: Sample assignment of an unequal number of bits for every value.

Residual Value	Code			Bit count
0	1	0		2
1	1	1	0	3
2	1	1	1	3
3	0			1

2.3.4   EXPLOITING PSYCHO-VISUAL REDUNDANCIES

Studies have shown that the human visual system (HVS) is much more sensitive to luminance information (Y) than chrominance information (U and V). This means that a reduction in the number of bits allocated for the chroma component will have a significantly lower impact on the visual experience than a corresponding reduction in luma bits.

By exploiting this visual perceptual reality, all modern video coding standards use a lower resolution of chroma by subsampling the chroma components, while maintaining full resolution of luminance components. The following are the most commonly used formats:

1. 4:2:0: Chroma subsampled by ½ across H and V directions
2. 4:2:2: Both U & V subsampled by ½ across H direction only
3. 4:4:4: Full resolution for U and V without any subsampling

The 4:2:2 and 4:2:0 subsampling mechanism is illustrated in Figure 13 for a sample 8x8 block of pixels, where full luma (Y) resolution is used but the chroma components (Cb and Cr) are both sampled as indicated by the shaded pixels. As 4:2:2 subsamples chroma along horizontal resolution only, every other pixel location along the horizontal rows is used. For 4:2:0, in addition to the above horizontal subsampling, vertical subsampling is also used. Hence, a pixel location that is between two consecutive rows of corresponding luma pixel locations is used.

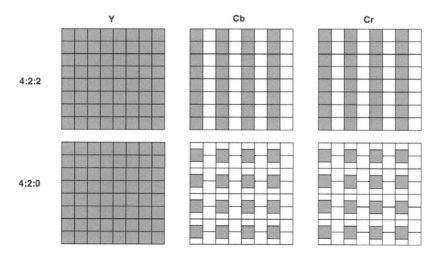

Figure 13: 4:2:2 and 4:2:0 subsampling of pixels.

Every video distribution system, for instance, satellite uplinks, cable or internet video, use 4:2:0 exclusively. In professional facilities like live event capture and news production centers, however, 4:2:2 format is used to capture, process, and encode video to preserve video fidelity.

Figure 14: HVS sensitivity to the low-frequency sky, and high-frequency trees, areas.

Figure 15: Lack of details are more prominent in large and smooth areas like the sky [3].

Additionally, the human eye is sensitive to small changes in luminance over a large area but not very sensitive to rapid luminance changes (high-frequency luminance). In the original image shown in Figure 14, the area around the sky corresponds to large, smooth, low-frequency areas, whereas the texture in the trees corresponds to high-frequency luminance areas.

When we begin to remove details in the picture as shown in Figure 15, the eyes notice the changes in the smoother areas like the sky, lake, and pathways more quickly than they do the changes in high-frequency grass and tree areas. The perceptual system is more tolerant of changes in the latter.

This is hugely important. What it means is that significant gains can be achieved by prioritizing low-frequency components over high-frequency components.

All video standards employ two important techniques to effectively achieve this.

1. **Transform Coding.** This is a process to convert luma and chroma components from the pixel domain to a different representation called the transform domain.
2. **Quantization.** Using this technique, low-frequency components are prioritized and preserved better. High-frequency components are selectively ignored.

2.3.5   8-BIT VERSUS 10-BIT ENCODING

It may seem obvious that using fewer bits to represent the video pixels will result in better compression. However, studies have shown that 10-bit encoding is capable of providing better quality over 8-bit encoding, regardless of the bit depth of the source content [4]. While this sounds counterintuitive, in the following I explain why it may be so.

If the content is 10-bit, then the filtering stages before encoding and after decoding could potentially destroy fine details with 8-bit encoding. However, these are preserved and leveraged in 10-bit encoding. On the other hand, if the source content is 8-bit but is encoded using 10-bits, the internal encoding process (like transforms and filters) uses at least 10-bit accuracy. This results in lesser rounding errors, especially in the motion compensated filtering process, thereby increasing the prediction efficiency. With more accurate prediction (which uses fewer bits), lower levels of quantization will be needed to achieve the target bitrate. The ultimate result of all this is superior visual quality.

2.4   SUMMARY

- Storage and transmission of raw uncompressed video consume enormous bandwidth (e.g. 3840x2160p60 takes 7.46 Gbits for one second). This is not practical, especially as video resolutions are increasing and consumption is growing. Hence, video needs to be compressed.
- Video pixels are similar with significant redundancies, classifiable into 2 types: statistical and psycho-visual. Compression techniques work to identify and remove these redundancies.
- Statistical redundancies are classified into spatial pixel redundancies (across pixels within a frame), temporal pixel redundancies (across frames) and coding redundancies.
- The human eye is more sensitive to luma than to chroma. Therefore, luma is prioritized over chroma in compression techniques.
- The human eye is sensitive to small changes in brightness over a large area but not very sensitive to rapid brightness variations. Therefore, low-frequency components are prioritized over high-frequency components during compression.

2.5   NOTES

1. Wong J I. The internet has been quietly rewired, and video is the reason why. Quartz Obsessions. https://qz.com/742474/how-streaming-video-changed-the-shape-of-the-internet/. Published October 31, 2016. Accessed September 21, 2018.
2. akiyo. xiph.org. Xiph.org Video Test Media [derf's collection]. https://media.xiph.org/video/derf/. Accessed September 21, 2018.
3. in_to_tree. xiph.org. Xiph.org Video Test Media [derf's collection]. https://media.xiph.org/video/derf/. Accessed September 21, 2018.
4. *Why Does 10-bit Save Bandwidth (Even When Content is 8-bit)?* ATEME. www.ateme.com. http://x264.nl/x264/10bit_02-ateme-why_does_10bit_save_bandwidth.pdf. Published 2010. Accessed September 21, 2018.

3 EVOLUTION OF CODECS

3.1 KEY BREAKTHROUGHS IN ENCODING RESEARCH

Video encoding technologies have consistently progressed over several decades since the 1980s with a suite of successful codecs built on the hybrid block-based architecture. The core technologies that constitute this architecture were developed over decades, starting from research as early as the 1940s. These underpinned the evolution of the architecture into its present form. Today's codecs still build on many significant research developments, especially from the 1970s and 1980s [1]. The focus of this section is to look at what these core technologies are and how they influenced and contributed to the evolution of the modern video coding standards. This provides us with valuable insights on why coding technologies are the way they are today. It also helps us understand, at a higher level, the fundamental framework of video coding.

3.1.1 INFORMATION THEORY (ENTROPY CODING)

It is widely accepted that the landmark breakthrough for communication systems that also laid the foundation of information theory was the publication of Claude Shannon's classic 1948 paper, " A Mathematical Theory of Communication."[2] In this paper, Shannon was able to provide a model for digital communication using statistical processes. He also introduced the concept of *entropy* to calculate the amount of *information* in a transmitted message, thereby enabling calculation of the limits of lossless data communication.

These concepts served as the building blocks for several entropy coding techniques, starting with David Huffman's 1952 paper, "A Method for the Construction of Minimum Redundancy Codes,"[3] that described a method of efficiently encoding a message with a finite set of symbols using variable length binary codes. The *Huffman coding method*, as we know it today, has been used extensively in video codecs, starting with H.261 and continuing up to H.264.

Witten's, et al. 1987 paper, "Arithmetic Coding for Data Compression"[4] provided an alternative to the Huffman coding method. Their technique improved compression efficiency and has formed the basis of CABAC encoding used in video coding standards, including H.264, H.265, and VP9.

3.1.2   PREDICTION

Given the nature of video signals, efforts to represent video data using some form of prediction, in order to minimize the redundancies and thereby reduce the amount of transmitted data, began as early as the 1950s. In 1972, Manfred Schroeder of Bell Labs obtained a patent, "Transform Coding of Image Difference Signals,"[5] that explored several of the modern video codec concepts, including inter-frame prediction, transforms and quantization to image signals. Schroeder's work also specifically mentions the application of Fourier, l-Hadamard, and other unitary matrix transforms that help to disperse the difference data homogeneously in the domain of the transformed variable.

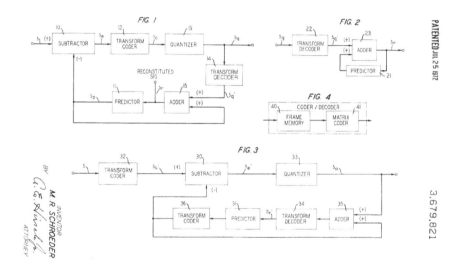

Figure 16: Manfred Schroeder's predictive coding techniques.
Image source: https://patents.google.com/patent/US3679821

While Schroeder's patent referenced inter frame prediction, it didn't specifically involve the modern-day concept of motion compensated prediction. This was introduced by Netravali and Stuller in their 1981 patent, "Motion Estimation and Encoding of Video Signals in the Transform Domain,"[6] that described the techniques of motion estimation and motion compensation for predictive coding. In this work, the first encoding step is a linear transform technique like Hadamard Transform, followed by motion compensated prediction that is carried out in the transform domain. These principles have since formed the backbone of all modern-day video codecs.

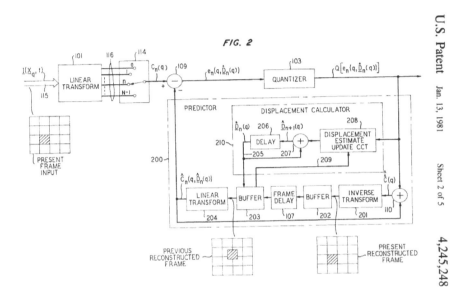

U.S. Patent   Jan. 13, 1981   Sheet 2 of 5   4,245,248

Figure 17: Netravali and Stuller's motion compensated prediction in the transform domain.
Source: https://patents.google.com/patent/US4245248A/

3.1.3   TRANSFORM CODING

The next important breakthrough was on achieving decorrelation of the source video (or image) pixels before processing and transmission. Prior work included references to various transform techniques to achieve this. In their 1974 paper, "Discrete Cosine Transform,"[7] Ahmed, Natarajan, and Rao introduced the DCT transform for image processing. As they explained, the DCT transform closely approximates the performance of the theoretically optimal Karhunen-Loeve Transform (KLT). The DCT and integer variants of the DCT have been adopted for transform coding in

MPEG standards, including MPEG1, MPEG2, MPEG4-Part2, H.264, and H.265.

Together, these developments and techniques have formed the pillars of the hybrid block-based video coding architecture used in today's codecs. This basic video architecture has been improved upon to produce newer codecs that deliver improved compression efficiency. How and when these coding tools were used in the development of various video coding standards is presented in the following section.

3.2 EVOLUTION OF VIDEO CODING STANDARDS

Most video codecs are part of an international collection of standards and it is useful to learn about how these standards were formed and how they have evolved over time. A video coding standard is a document that describes the video bitstream format and an associated mechanism to decode the bitstream. It only defines the decoder. It leaves the encoder unspecified, allowing flexibility in encoder research and implementation to provide the compressed video in the prescribed format. Included in the decoder definition are also coding tools that can be used for creating the compliant bitstream. This document is usually approved and adopted by an international standards body like ISO and/or ITU-T.

These bodies have done this successively to produce a host of MPEG video coding standards. The standardization process ensures that there is a fixed reference that can be used by designers and manufacturers of encoders and decoders to make the devices interoperable. This means encoders made by one manufacturer will produce bitstreams that can be decoded by decoders made by any other manufacturer. This also allows freedom for the consumer to be able to choose devices from different manufacturers, making the market highly competitive.

Over the years, the process of standardization has also become fairly standard. Once the requirements, including target applications and bitrates, are understood, a development process ensues where technical algorithm contributions are sought from individuals or corporations. When available, these are analyzed competitively for performance against a set of criteria and some are selected for finalization. The draft standard document is then generated. It includes these selected algorithms and techniques that, upon successful compliance testing, evolve into the final standard.

While ISO and ITU-T have standardized and published several video coding standards, including all the MPEG and H.26x video standards, a parallel mechanism, spearheaded by technology companies like Google and others, has produced other video standards like VP8 and VP9. This mechanism is similar to the one used by traditional standards bodies to develop and test new algorithms. However, the standard document and software are usually published by corporations like Microsoft and Google. The standard may or may not then be adopted and published by a standards body. (Microsoft's VC-1 codec was adopted as a SMPTE standard.)

Some of these video codecs, like VP8 and VP9, are also available as free open-source resources to the public. This mechanism has also evolved over the years, resulting in the formation of an alliance of major companies that contribute to the development and shared usage of the coding standard and resources. The following section provides insights on the timeline and development of popular video coding formats.

3.2.1   TIMELINES AND DEVELOPMENTS

Two major standards bodies, namely, ISO/IEC and ITU-T, have, over the years, published the majority of video coding standards. The Motion Pictures Experts Group (MPEG) has been the working group for ISO/IEC standards and the Video Coding Experts Group (VCEG) has been the working group under the ITU-T. While the ISO/IEC produced the popular MPEG series of standards, ITU-T produced, through separate efforts, the competing H.26x series of standards.

This arrangement, however, changed with the formation of the Joint Video Team (JVT), a collaboration between VCEG and MPEG that worked together and developed the H.264 or MPEG-4 Part-10 video coding standard. It was an important milestone as it brought together several coding tools from earlier H.26x and MPEG standards.

The High Efficiency Video Coding (HEVC or H.265 or MPEG-H Part2) standard, that builds upon H.264, is also published by the Joint Collaborative Team on Video Coding (JCT-VC), a collaborative effort by the same groups.

3.2.1.1   MPEG VIDEO STANDARDS

MPEG1 Video was standardized in 1993 and was used for digital video storage on VCDs with a bit rate target of 1.5 Mbps for VHS quality video. It included support for CIF resolution (352x288 pixels) with YUV 420. This was extended to MPEG2 video which was standardized in 1995 and was the first commercially successful codec for broadcast applications targeting high bit rates of 3-20 Mbps and included HD video resolutions. MPEG2 revolutionized video transmission and storage and also pushed the evolution of digital television as we know it today. It was also used as the standard of storage on DVDs and the standard is still extensively in use today. MPEG4 Part 2 was the next standard that was built on the same principles as MPEG2, with additional coding tools. It targeted low bit rate applications like web streaming and video calling. It also included compatibility with ITU-T's H.263 standard. It had limited success and was soon replaced by the next generation MPEG4-Part 10 or H.264 standard.

3.2.1.2   H.26X VIDEO STANDARDS

In parallel with the development of the MPEG standards, the VCEG, under the ITU-T, published the H.26x standards, including H.261, H.262, and H.263. H.261 predated MPEG1 and was standardized in 1988. It supported CIF (352x288) and QCIF (176x144) resolutions with YUV 4:2:0 format for low latency video conferencing applications. H.262 was standardized as the MPEG2 video standard, establishing the collaboration between the two standards bodies. H.263 was standardized in 1996 and built upon H.261 to provide enhancements for video conferencing applications.

3.2.1.3   JOINT COLLABORATION STANDARDS

With the promise to reduce the bit rate further by 50% over MPEG2 while maintaining the same video quality, the H.264 or AVC video standard was jointly developed in the early 2000s and standardized in 2003 by the JVT. This was hugely successful and has been deployed widely for a variety of applications, ranging from linear broadcasting, internet video streaming and storage applications like Blu-Ray Disks. It has become the de facto codec of the internet.

The success of H.264 led to the same model for the development of H.265 or HEVC video standard that was standardized in 2013. While HEVC retained the basic structure of H.264, it added significant improvements that resulted in a reduction of the bitrate of 50% while maintaining comparable video quality. The high efficiency coding, however, came at the expense of significantly high complexity of algorithms for both encoder and decoder.

The next video codec, called VVC (Versatile Video Coding), is being developed by The Joint Video Experts Team (JVET) with a goal of decreasing the bit rate by a further 50% over H.265. This will provide rich improvements to encode 4K, 8K and even higher resolutions and will also target applications like 360-degree and high-dynamic-range (HDR) video. With the first draft expected by late 2019, the standard is expected to be ready somewhere in the timeframe of late 2020 to early 2021.

3.2.1.4 OPEN SOURCE VIDEO FORMATS

While MPEG and H.26x video standards have had the lion's share of development, implementation and deployment thus far, other popular video formats have also evolved to prominence over the years, especially as video transmission is transitioning from traditional methods to internet processing and delivery. These include Google's VP8, VP9, and the emerging AV1 standard.

VP8 was originally developed by On2 Technologies and later acquired by Google and released as an open and royalty free codec in 2010.

Table 5: Timelines of the evolution of video coding standards.

Year	Standard	Applications
1988	H.261	Video conferencing
1992	MPEG1	Storage (VCDs)
1995	H.262/MPEG2	Storage (DVD) & broadcast transmission
1996	H.263	Low bitrate video conferencing
1999	MPEG4-Part2	Low bitrate applications like web streaming

Year	Standard	Applications
2003	H.264/MPEG4-Part 10	Video streaming, storage and broadcast transmission
2008	VP8	Internet video streaming
2013	H.265/MPEGH-Part 2	Video streaming, storage and broadcast transmission
2013	VP9	Internet video streaming
2018	AV1	Internet video streaming

VP9, the successor to VP8 was developed by Google and expands on VP8's coding tools. It was primarily targeted at YouTube streaming but has since been expanded to other internet streaming platforms, including Netflix. Google has since also spearheaded the drive for royalty-free video coding formats by forming the Alliance for Open Media (AOM) along with several firms in the video industry, including Netflix, Mozilla, Cisco, and others. The alliance's first joint codec, AV1, is released in 2018 and, while largely based on VP9, has been built on Xiph's/Mozilla's Daala, Cisco's Thor and Google's own VP10 coding formats.

Over the course of this evolution, every generation has built on top of the previous generation by introducing new toolsets that focus primarily on reduction of bit rates, reduction of decoder complexity, support for increased resolutions, newer technologies like multi-view coding, HDR, and improvements in error resilience, among other enhancements. Table 5 consolidates the details of the evolution of video coding standards.

3.2.2   COMPARISON TABLE OF MPEG2, H.264, AND H.265

Along with the timelines of development and target applications of video codecs, it's also important to review the constituent tool sets that every generation of codec added to deliver compression efficiency. We will explore the evolution of coding toolsets by comparing three popular video coding standards. This will provide a representative overview of this technological evolution. The details of various coding tools will be covered in subsequent chapters.

Thus, even if the following terms may not mean much at this point, the summary in Table 6 can serve to provide a broad overview of the nuts and bolts that constitute every modern codec.

Table 6: Comparison of toolsets in modern video coding standards.

Coding Tool	MPEG 2	H.264	H.265
Block size	macroblock of 16x16	macroblock of 16x16	variable 8x8 to 64x64 CTUs
Partitioning	16x16	variable partitions from 16x16 down to 4x4	variable partitions from 64x64 down to 4x4
Transforms	floating point DCT based 8x8 transforms	4x4 and 8x8 integer DCT transforms	variable 32x32 down to 4x4 integer DCT transforms + 4x4 integer DST transform
Intra prediction	DC prediction in the transform domain	spatial pixel prediction with 9 directions	spatial pixel prediction with 35 directions
Sub pixel interpolation	½ pixel bilinear filter	½ pixel six-tap filter and ¼ pixel bilinear filter	¼ pixel eight-tap filter Y and ⅛ pixel four-tap UV
Filtering	no in-loop filtering	in-loop deblocking filter	in-loop deblocking and SAO filter
Entropy coding	VLC	CAVLC and CABAC	CABAC
Block skip	none	direct modes	Merge modes

Coding Tool	MPEG 2	H.264	H.265
modes			
Motion vector prediction	spatial MV prediction from one neighbor	spatial prediction using 3 neighboring MVs	Enhanced spatial and temporal prediction
Parallelism tools	slices	slices and tiles	Wavefront parallel processing, tiles, slices
Reference pictures	2 reference pictures	up to 16 depending on resolutions	up to 16 depending on resolutions
Interlaced coding	Field and frame coding are supported.	Field, frame, and MBAFF modes are supported.	only frame coding is supported.

3.3   SUMMARY

- The core technologies that constitute modern compression architecture were developed over decades of research starting in the 1940s.
- The three key breakthroughs that propelled video compression systems in their present form are a) information theory, b) prediction, and c) transform.
- Two major standards bodies, namely, ISO/IEC and ITU-T, have, over the years, published the majority of video coding standards. The Motion Pictures Experts Group (MPEG) has been the working group for ISO/IEC standards and the Video Coding Experts Group (VCEG) has been the working group under the ITU-T.
- MPEG and VCEG have collaborated to jointly produce immensely successful video standards, including MPEG2, H.264, and H.265.

- Google published VP9 as an open video coding standard in 2013. Its success has led to the formation of an alliance called AOM that is working on a new standard called AV1.
- The Joint Video Experts Group (JVET) is working on the successor to H.265, called VVC (Versatile Video Codec), with a goal of decreasing the bit rate by a further 50% over H.265.

3.4   NOTES

1. Richardson I, Bhat A. Video coding history Part 1. Vcodex. https://www.vcodex.com/video-coding-history-part-1/. Accessed September 21, 2018.
2. Shannon CE. A mathematical theory of communication. Bell Syst Tech J. 1948;27(Jul):379-423;(Oct):623-656. https://goo.gl/dZbahv. Accessed September 21, 2018.
3. Huffman DA. A method for the construction of minimum-redundancy codes. Proc IRE. 1952;40(9):1098-1101. https://goo.gl/eMYVd5. Accessed September 21, 2018.
4. Witten IH, Neal RM, Cleary JG. Arithmetic coding for data compression. Commun ACM. 1987;30(6):520-540. https://goo.gl/gRrXaS. Accessed September 21, 2018.
5. Schroeder MR. Transform coding of image difference signals. Patent US3679821A. 1972.
6. Netravali AN, Stuller JA. Motion estimation and encoding of video signals in the transform domain. Patent US4245248A. 1981.

Ahmed N, Natarajan T, Rao KR. Discrete cosine transform. IEEE Trans Comput. 1974;23(1):90-93. https://dl.acm.org/citation.cfm?id=1309385. Accessed September 21, 2018.

PART II

4 VIDEO CODEC ARCHITECTURE

Video compression (or video coding) is the process of converting digital video into a format that takes up less capacity, thereby becoming efficient to store and transmit. As we have seen in Chapter 2, raw digital video needs a considerable number of bits and compression is essential for applications such as internet video streaming, digital television, video storage on Blu-ray and DVD disks, video chats, and conferencing applications like FaceTime and Skype.

Compressing video involves two complementary components. At the transmitting end, an *encoder* component converts the input uncompressed video to a compressed stream. At the receiving end, there is a *decoder* component that receives the compressed video and converts it back into an uncompressed format.

The word 'codec' is derived from the two words - encode and decode.

CODEC = En**co**de + ***Dec***ode

Quite obviously, there are numerous ways in which video data can be compressed and it therefore becomes important to standardize this process. Standardization ensures that encoded video from different sources using products of different manufacturers can be decoded uniformly across products and platforms provided by other manufacturers.

For example, video encoded and transmitted using iPhone needs to be viewable on an iPhone and on a Samsung tablet. Streamed video from Netflix or YouTube needs to be viewable on a host of end devices. It needs no further emphasis that this interoperability is critical to mass adoption of the compression technology.

All modern video coding standards, including H.264, H.265 and VP9, define a bitstream syntax for the compressed video along with a process to decode this syntax to get a displayable video. This is referred as the *normative section* of the video standard. The video standard encompasses all the coding tools, and restrictions on their use, that can be used in the standard to encode the video. The standard, however, does not specify a process to encode the video.

While this provides immense opportunity for individuals, universities and companies for research to come up with the best possible encoding schemes, it also ensures that every single encoded bitstream adhering to the standard can be completely decoded and produce identical output from a compliant decoder. Figure 18 shows the encoding and decoding processes and the shaded portion in the decoder section highlights the normative part that is covered by video coding standards.

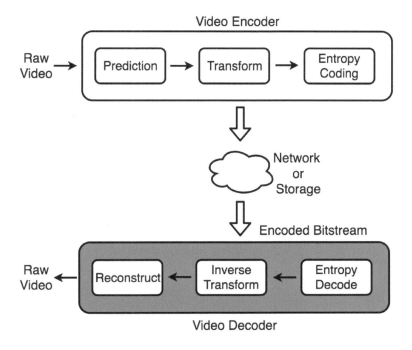

Figure 18: The process of encoding and decoding.

4.1   HYBRID VIDEO CODING ARCHITECTURE

As digital video is represented by a sequence of still images, it's natural that video compression technologies employ frameworks to analyze and compress every picture individually. To do this, every frame is categorized either as an intra or an inter frame and different techniques are used to identify and eliminate the spatial and temporal redundancies to achieve efficient compression. In the first part of this section, we shall explore how individual pictures in the video sequence are categorized and grouped together for encoding. The latter half of this section then goes into how the frames themselves are broken down further for predictive block-based encoding.

4.1.1   INTRA FRAME ENCODING

Intra frame encoding uses only the information in the current frame by analyzing and removing the spatial correlation amongst pixels to minimize the frame size. An intra frame, or *I frame*, is thus a self-contained frame that can be independently encoded and correspondingly decoded without any dependency on other frames.

It's not uncommon to find I frames with a compression ratio around 1:10, meaning that the size of an I frame is 10 times lower than its uncompressed version. However, the actual compression can vary depending on the bit rate, coding tools used, and settings used to encode the video.

As I frames don't reduce temporal redundancies but only exploit pixel redundancies within the same frame, the drawback of using these frames is that they consume many more bits. On the other hand, owing to their lower compression ratio, they do not generate many artifacts and hence serve as excellent reference frames to encode subsequent, temporally predicted frames.

Intra frame encoding is less computationally expensive than inter frame encoding and also doesn't require multiple reference frames to be stored in memory.

The first frame in a video sequence is encoded as an I frame. In live video transmission, these are used as starting points for newer clients (viewers) to join the stream. They are also used as points of resynchronization when the decoder encounters transmission errors in the bitstream. In compressed storage devices like DVD and Blu-Ray discs, I frames are used to implement random access points and *trick modes* like fast-forward and rewind.

4.1.2   INTER FRAME ENCODING

In a typical video sequence, the individual pictures are captured and played back at typical rates of either 25, 30 or up to 60 frames in one second. Unless the section of the visual sequence has a complete scene change or high motion, it's likely that subsequent frames of the video are similar, with the same objects in more or less similar positions. This is true, for example, of a news or a chat show that has talking heads with static background. In such scenes, it's more efficient to focus on and analyze the *changes* between the pictures rather than analyzing the actual pictures themselves. That's what inter frame encoding is all about.

In addition to analyzing the current frame, inter frame compression also analyzes the information from neighboring frames and uses difference coding to remove the temporal, inter-picture redundancies to achieve compression. In difference coding, a frame is compared with an earlier encoded frame that is used as a reference frame. The difference between their pixel values, called *residual*, is then calculated.

Only the residual is encoded in the bitstream. This ensures that only those pixels that have changed with respect to the reference frame are coded. In this way, the amount of data that is coded and sent is significantly reduced. Quite possibly, as noted earlier, the vast majority of the scene hardly changes between pictures (unless there is a scene change or significant motion). Thus, this method usually leads to a significant reduction in the number of bits needed to encode the frame.

Inter frame encoding uses a variety of techniques like *motion estimation* and *motion compensation* to encode the changes from one frame to the next.

Motion estimation is a process that analyzes different frames and provides motion vectors that are used to describe the motions of various objects across these frames. By incorporating this motion vector information in difference coding, coding efficiency is significantly improved, especially if the video scene contains several moving objects. The process of motion estimation is further explained in detail in chapter 6.

Inter frame encoding can employ the following two types of predictive coding.

4.1.2.1 PREDICTIVE CODED FRAMES (P FRAMES)

P frame (predictive inter frame) encoding uses frames that were encoded earlier in the video sequence as a reference and encodes the changes in pixel values of the current frame from the pixels in the reference frame. As illustrated in Figure 19, below, the current frame (P frame) and the reference frame (I frame) have similarities and thus prediction using the I frame as a reference helps in reducing the number of encoded bits.

The reference frames are usually I or P frames, however, newer codecs also use other frames like *B frames* (bidirectionally predictive inter frames) for reference. While reliance on the previous frames helps in reducing the number of bits used to encode the inter frames it usually increases the sensitivity to transmission losses. As an example, if there's a transmission error that results in loss of bits in an I frame, a bitstream with inter frames that reference the I frame will show more visual artifacts than if the bitstream were encoded with only I frames.

A P frame is thus the term used to define *forward prediction* and consists of motion information and residual data that effectively determine the prediction.

4.1.2.2 BIDIRECTIONALLY PREDICTIVE CODED FRAMES (B FRAMES)

B frames are similar to P frames, except that they reference and extract information from frames that are temporally before and later in the sequence. Thus, the difference between P frames and B frames lies in the type of reference frames they use.

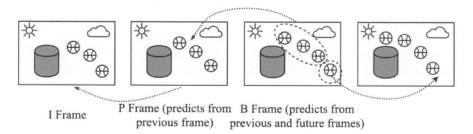

I Frame   P Frame (predicts from   B Frame (predicts from
                 previous frame)   previous and future frames)

Figure 19: Illustration of P frame and B frame encoding.

To understand why this is needed and also the benefits that this type of prediction offers, let us examine the sequence of frames shown in Figure

19. If the third frame in the sequence were to be encoded as a P frame that uses only forward prediction, it would lead to poorer prediction for the ball region, as the additional ball is not present in the preceding frames. However, compression efficiency can be improved for such regions by using backward prediction. This is because the ball region can be predicted from the future frames that have the fourth moving ball in them. Thus, as shown in the figure, the third frame could benefit from B frame compression by having the flexibility to choose either forward or backward prediction for the region containing the top three balls and any of the future frames for the last ball.

To be able to predict from future frames, the encoder has to, *in an out-of-order fashion*, encode the future frame before it can use it as a reference to encode the B frame. This requires an additional buffer in memory to temporarily store the B frame, pick and encode the future frame first and then come back to encode the stored B frame. Because the frames are sent in the order in which they are encoded, in the encoded bitstream, they are available *out-of-order* relative to the source stream. Correspondingly, the decoder would decode the *future frame* first, store it in memory and use it to decode the B frame. However, the decoder will sequentially display the B frame first, followed by the I frame. Consequently, as a B frame is based on a frame that will be displayed in the future, there will be a difference between decode order and display order of frames in the sequence when the decoder encounters a B frame.

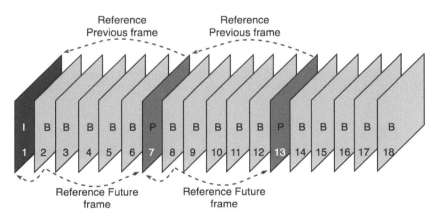

Figure 20: Sequence of frames in display order as they appear in the input.

We can see this clearly in Figure 20, above. It shows the display order of frames in the video sequence. This is also the same sequence of frames as they appear in the input. This display number, or frame number, is also indicated in the frames in the figure. In this example, I, P, and B frame encoding is used and there are five B frames for every P frame. Frame number 1 is encoded as an I frame, followed by a future frame (frame number 7) as a P frame, so that the five B frames can have future frames from which they can predict. After frame 7 is encoded, frames 2 to 6 are also encoded. This pattern is repeated periodically, with frame 13 encoded earlier as a P frame, and so on.

The encoding (and corresponding decoding) order of the frames, as they appear in the bitstream for this specific example, is shown in Figure 21. During the decoding process the P frames (frames 7, 13, 19) have to be decoded before the B frames (frames 2-6, 8-12, 14-17) that precede them can be decoded. However, the P frames will be held in a buffer and displayed only after the B frame is displayed.

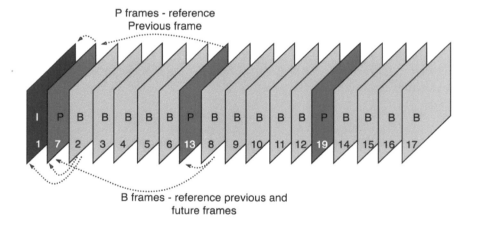

Figure 21: Sequence of frames in encode/decode order as they appear in the bitstream.

Typically, B frames use far fewer coding bits than P frames and there are many B frames encoded for every P frame in the sequence. A B frame is thus the term used to define both forward prediction and backward prediction and consists of motion vectors and residual data that effectively describe the prediction. As with P frames, the reliance on the previous and future frames helps in further reducing the number of bits used to encode the B frames. However, it increases the sensitivity to transmission losses. Furthermore, the cumulative prediction and quantization (a term that will be explained in chapter 7) processes across successive frames increase the error between the original picture and the reconstructed picture. This is because quantization is a lossy process and prediction from a lossy version of the picture results in increased error between the original and reconstructed pictures. For this reason, earlier standards did not use B frames as reference frames. However, the enhancements in filtering and prediction tools have improved the prediction efficiency, thereby enabling the use of B frames as reference frames in newer standards like H.264, H.265, and VP9.

The relative frame sizes by each of these frame types are illustrated in Figure 22. The figure is a graph plotting the file sizes of a sample file encoded with I, P, and B frame types using an I frame period of 50 frames. In this graph, the peaks are the frame sizes of I frames that occur at every interval of 50 frames. These are represented in red color. The next largest frame sizes are P frames represented in blue color followed by B frames which are shown in green color.

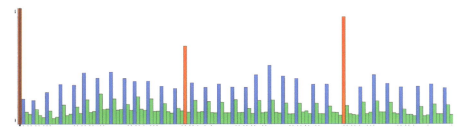

Figure 22: Illustration of frame size across different frame types.

4.1.3   GROUP OF PICTURES (GOP) STRUCTURES

Every frame type has its role to play in an encoded sequence. While I frames consume more bits they serve as excellent references and access points in the bitstream. P and B frames leverage their predictions to crank up compression efficiency.

Most video sequences have similar images for long periods of time. By strategically interspersing I, P and B frames periodically, such that there are many P and B frames between each I frame, it is possible to obtain dramatically higher compression, on the order of 1000:1, while maintaining acceptable visual quality levels.

The sequence of periodic and structured organization of I, P, and B frames in the encoded stream is called a group of pictures (GOP). A GOP starts with an I frame. This allows fast seek and random access through the sequence. Upon encountering a new GOP (I frame), the decoder understands that it doesn't need any information from previous frames for further decoding and resets its internal buffers. Decoding can thereby start cleanly at a GOP boundary and prediction errors from the previous GOP are corrected and not propagated further.

4.1.3.1 OPEN AND CLOSED GOPS

For the last pictures in the GOP, the encoder also has the option to use frames from the subsequent GOP for bidirectional prediction. By doing so, better prediction and, hence, slightly better video quality can be achieved. The resulting GOP using this mechanism is called an open GOP. Alternatively, the encoder can choose not to rely on frames from subsequent GOPs and instead keep the prediction for all frames contained within the same GOP. This mechanism is called a closed GOP. As a closed GOP is self-contained without any neighboring GOP dependencies, it is useful for frame-accurate editing and also serves as a good splice point in the bitstream.

GOP intervals are expressed in seconds. A 1s GOP means the encoder inserts an I frame at every interval of 1 second. This means that if the frame rate of the video is 30fps (say 1080p30), the encoder inserts an I frame for every 30 frames. However, if the video frame rate is 60fps, then it inserts an I frame every 60 frames.

The inherent GOP structure is often represented by two numbers, namely, M and N. The first number, M, refers to the distance between two reference P frames and the second number, N, refers to the actual GOP size. Modern codecs may use the count of B frames instead of M. As an example, M=5, N=15, indicates the GOP structure illustrated in Figure 23. Here, every 5th frame is a reference P frame and an I frame follows after every 15 frames.

I BBBBP BBBBP BBBB | I BBBBP BBBBP BBBB | I ...

Figure 23: Illustration of GOP with M=5 and N=15.

Modern encoders have great flexibility in choosing from among a host of schemes for reference frames, especially as B frames may also be used as references to code other (B or P) frames. This helps to significantly improve compression but is sensitive to error propagation. This means that if some data gets lost, the complex referencing structures will serve to propagate the errors introduced.

Hierarchical B reference schemes (also called B-pyramid referencing) that were introduced in H.264 and also supported in H.265 provide very good compression efficiency and can also limit the error propagation. The hierarchy that exists in referencing B frames helps to limit the number of pictures affected by data corruption. Let us explore a few typical GOP structures that modern encoders deploy.

1. **IBBBBP Structure:** This is the classic I-P-B GOP structure without any hierarchical B reference structures. The reference pictures, display and sample encode/decode order for every frame are explained in Figure 24. The P picture with sequence number 4 is encoded after the first I picture. This is then followed by the three B pictures in sequence, and so on. It should be noted that B pictures are still used for referencing other B pictures, without any hierarchy.

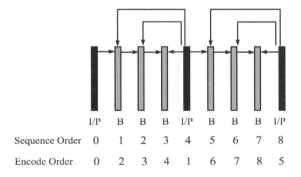

	I/P	B	B	B	I/P	B	B	B	I/P
Sequence Order	0	1	2	3	4	5	6	7	8
Encode Order	0	2	3	4	1	6	7	8	5

Figure 24: P and B reference frames without hierarchical B prediction.

2. **IBBBBBBBP Structure:** This scheme uses 7 B frames, some of which are used as reference frames for other P and B frames as shown in Figure 25. Frame numbers 0 and 8 in the sequence or display order that are I/P are encoded first, followed by frame 4 (B1). Frame 4 uses I and P as reference frames. B1 frames are the first level in the hierarchy and are also used as references for the next level B frames, namely, B2. This scheme further

extends hierarchically to frames in level B3 that use B frames in level B2 as reference.

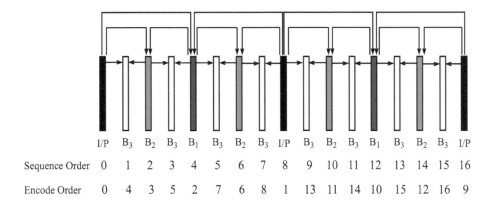

	I/P	B$_3$	B$_2$	B$_3$	B$_1$	B$_3$	B$_2$	B$_3$	I/P	B$_3$	B$_2$	B$_3$	B$_1$	B$_3$	B$_2$	B$_3$	I/P
Sequence Order	0	1	2	3	4	5	6	7	8	9	10	11	12	13	14	15	16
Encode Order	0	4	3	5	2	7	6	8	1	13	11	14	10	15	12	16	9

Figure 25: Hierarchical B reference frames.

4.2   BLOCK-BASED PREDICTION

Modern codecs like H.265 and VP9 employ a hybrid, block-based prediction architecture. This is the encoder design that we will deal with throughout this book. The block diagram of such a hybrid video encoder, adapted from Sullivan, Ohm, Han, & Wiegand, [1] is illustrated in Figure 26. The encoder uses a raw video sequence as input to create a standard compliant encoded bitstream.

In this model, each picture of the video sequence is categorized as an intra or inter picture for the purposes of prediction. The first picture in the sequence is marked as intra. This picture uses only intra prediction and thereby has no coding dependency on any other pictures. Because intra pictures consume more bits compared to inter pictures, they are used sparingly. The periodic interval in which they are inserted in to the bitstream is called the I frame interval. As we know, this also provides a random-access point into the video sequence.

Furthermore, during scene changes in the sequence, the encoder usually has built-in scene change detection algorithms and inserts an I frame at the scene change picture. Other pictures are coded as inter frames. These use temporal prediction from neighboring pictures and blocks.

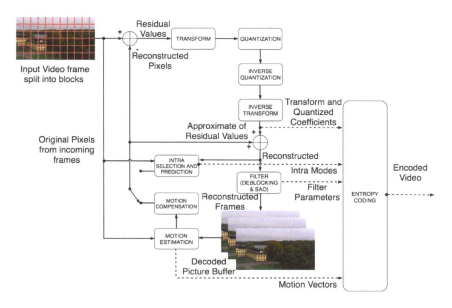

Figure 26: Block diagram of a block-based encoder.

The first step in every block-based encoder is to split every frame into block-shaped regions. These regions are known by different names in different standards. H.264 standard uses 16x16 blocks called *macroblocks*, VP9 uses 64x64 blocks called *superblocks* and H.265 uses a variety of square block sizes called *coded tree units* (CTUs) that can range from 64x64 to 16x16 pixels. With the increased need for higher resolutions over the years, the standards have evolved to support larger block sizes for better compression efficiency. The next generation codec, AV1, also supports block sizes of 128x128 pixels. Every block in turn is usually processed in raster order within the frame in the encoding pipeline. These blocks are further broken down in a recursive fashion and the resulting sub blocks are then processed independently for the prediction.

Figure 27 shows how the recursive partitioning is implemented in VP9. Each 64x64 superblock is broken down in either of the 4 modes, namely, 64x32 horizontal split, 32x64 vertical split, 32x32 horizontal and vertical split mode, or, no split mode.

Recursive splitting is permitted in the 32x32 horizontal and vertical split mode. In this mode, each of the 32x32 blocks can be again broken down into any of the 4 modes and this continues until the smallest partition size is 4x4. This type of splitting is also called *quadtree split*. Figure 27 also illustrates how a 64x64 superblock in VP9 is broken down in a recursive manner to different partition sizes.

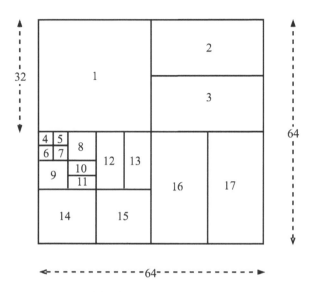

Figure 27: A 64x64 superblock in VP9 is partitioned recursively in to sub partitions.

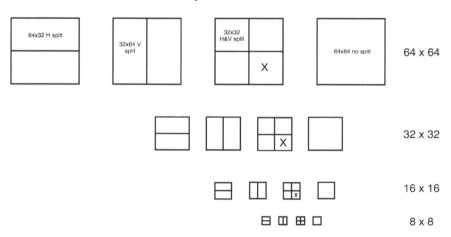

Figure 28: Recursive partitioning from 64x64 blocks down to 4x4 blocks.

At the highest level, the superblock is broken down using split mode to four 32x32 blocks. The first 32x32 block is in *none mode.* This is not broken down again. The second 32x32 block has horizontal mode. This is split into two partitions of 32x16 pixels, each (indicated as 2 and 3). The third 32x32 block is in split mode; hence, again, recursively broken in to four 16x16 blocks. These are further broken down all the way to 8x8 and further down to 4x4 blocks. This mechanism of recursive partition split is also illustrated in Figure 28. The partition scheme in HEVC is very similar with a few minor variations.

Now that we understand the recursive block sub partitioning schema, let us explore why this kind of partition is needed and how it helps.

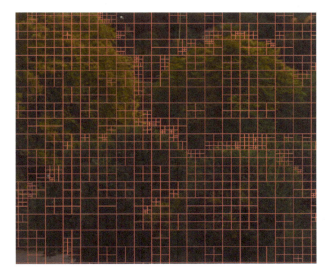

Figure 29: Partition of a picture in to blocks and sub partition of blocks.
Analyzer source: https://www.twoorioles.com/vp9-analyzer/

Coding standards like HEVC and VP9 address a variety of video resolutions from mobile (e.g., 320x240 pixels) to UHD (3840x2160 pixels) and beyond. Video scenes are complex and different areas in the same picture can be similar to other neighboring areas. If the scene has a lot of detail with different objects or texture, it's likely that smaller blocks of pixels are similar to other, smaller, neighboring blocks or also to other corresponding blocks in neighboring pictures.

Thus, we see in Figure 29, that the area with fine details of the leaves has smaller partitions. The interpixel dependencies in the smaller partition areas can be exploited better using prediction at the sub-block level to get better compression efficiency. This benefit, however, comes with increased cost in signaling the partition modes in the bitstream. Flat areas, on the other hand, like the darker backgrounds with few details, will have good prediction even with larger partitions.

The next question that comes to mind is, how are these partitions determined? The challenge to any encoder is to use the partition that best enables encoding of the pixels using the fewest bits and yields the best visual quality. This is usually an algorithm that is unique to every encoder. The encoder evaluates various partition combinations against set criteria and picks one that it expects will require the fewest encoding bits. The information on what partitioning is done for the superblock or CTU block is also signaled in the bitstream.

Once the partitions are determined, the following steps are done in every encoder in sequence. At first, every block undergoes a prediction process to remove correlation among pixels in the block. The prediction can be either within the same picture or across several pictures. This involves finding the best matching prediction block whose pixel values are subtracted from the current block pixels to derive the *residuals*. If the best prediction block is from the same picture as the current block, then it is classified as using intra prediction.

Otherwise, the block is classified as an inter prediction block. Inter prediction uses motion information, which is a combination of motion vector (MV) and its corresponding reference picture. This motion information and the selected prediction mode data are transmitted in the bitstream. As explained earlier, blocks are partitioned in a recursive manner for the best prediction candidates. Prediction parameters like prediction mode used, reference frames chosen, and motion vectors can be specified, usually for each 8x8 block within the superblock.

The difference between the original block and the resulting prediction block, called the *residual block*, then undergoes transform processing using a spatial transform. Transform is a process that takes in the block of residual values and produces a more efficient representation. The transformed pixels are said to be in *transform domain* and the coefficients are concentrated around the top left corner of the block with reducing values as we traverse the block horizontally rightward and vertically downward. Until this point, the entire process is lossless. This means, given the transformed block, the original set of pixels can be generated by using an inverse transform and reverse prediction.

The transform block then undergoes a process called *quantization* that involves dividing the block values by a fixed number to reduce the number of residual coefficients. These can then be efficiently arranged using a scanning process and then encoded to produce a binary bitstream using an entropy coding scheme.

When the decoder receives this video bitstream, it carries out the complementary processes of entropy decoding, de-scanning, de-quantization, inverse transform and inverse prediction to produce the decoded raw video sequence. When B frames are present in the stream, the decoding order (that is, bitstream order) of pictures is different from the output order (that is, display order). When this happens, the decoder has to buffer the pictures in its internal memory until they can be displayed.

It should be noted that the encoder does not employ the input source material for its prediction process. Instead, it has a built-in decoder processing loop. This is needed so that the encoder can produce a prediction result that is identical to that of the decoder, given that the decoder has access only to the reconstructed pixels derived from the encoded material. To this end, the encoder performs inverse scaling and inverse transform to duplicate the residual signal that the decoder would arrive at. The residual is then added to the prediction and loop-filtered to arrive at the final reconstructed picture. This is stored in the decoded picture buffer for subsequent prediction. This exactly matches the process and output of the decoder and prevents any pixel value drift between the encoder and the decoder.

4.3 SLICES AND TILES

We have seen in earlier sections how a frame can be split in to contiguous square blocks of pixels called superblocks or CTUs for processing. Modern codecs also employ *slices* and *tiles*. These split the frame in to regions containing one or more superblocks or CTUs. In H.264 and H.265, slices and tiles are used to split the frame into multiple units that are independently processed either in raster order (slices) or non-raster order (tiles) so that the encode/decode of these independent units can happen in parallel. This serves to speed up the computations in parallel architectures and multi-threaded environments. Tiles are also supported in VP9 and AV1 standards and are a computation-friendly toolset, especially for software, CPU-based encoder and decoder implementations.

The slice tool set is available in H.264 and H.265. These split every picture into independent entities containing blocks of the frame in raster order. In Figure 30, the frame is split into three slices, each containing several CTU blocks and every slice can be independently encoded. The CTUs are encoded in raster order within each slice.

This also allows the slices to be decoded independently, with the exception of *deblocking filtering operation*. This is permitted across multiple slice boundaries prior to reconstruction of the pixels for further prediction.

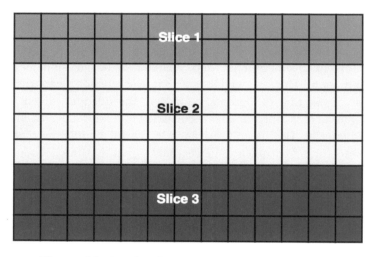

Figure 30: A video frame is split in to three slices.

VP9 supports tiles and, when implemented, the picture is broken along superblock boundaries. Each tile contains multiple superblocks that are all processed in raster order and flexible ordering is not permitted. However, the tiles themselves can be in any order. This means that the ordering of the superblocks in the picture is dependent on the tile layout. This is illustrated in Figure 31 below. It should be noted that tiles and slices are parallelism features intended to speed up processing. They are not quality improvement functions. This means that, in order to achieve parallel operations, some operations like predictions, context sharing, and so on would not be permitted across slices or tiles. This is to facilitate independent processing. Such limitations may also lead to some reduction in compression efficiency.

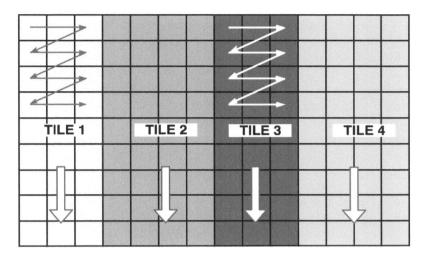

Figure 31: Splitting a frame in to 4 independent column tiles.

As an example, VP9 imposes the restriction that there can be no coding dependencies across column tiles. This means that two column tiles can be independently coded and hence decoded. For example, in a frame split into 4 vertical tiles, as shown in Figure 31, above, there will be no coding dependencies like motion vector prediction across the vertical tiles. Software decoders can therefore use four independent threads, one each to decode one full column tile. The tile size information is transmitted in the picture header for every tile except the last one. Decoder implementations can use this information to skip ahead and start decoding the next tile in a separate thread. Encoder implementations can use the same approach and process superblocks in parallel.

4.4   INTERLACED VERSUS PROGRESSIVE SCAN

Earlier when video was represented using analog technologies, it was captured and displayed by scanning alternate rows at two different times to enhance motion perception. Each of these rows was called a *field* and the resulting video was called *interlaced* video. This method of representing video has become lesser used in recent times. This is driven by the fact that digital displays today largely support progressive formats. With the widespread use of video over the internet, newer standards like H.265 and VP9 expect progressive content. There is no explicit coding tool in the recent standards to support interlaced coding. Video streaming and delivery over the internet is all progressive, further lessening the need to support the legacy interlaced coding. In this book, we will also focus exclusively on progressive scanned coding.

The coding tools and frameworks described above help to achieve the best compression efficiency. This is the goal of any video compression standard. However, standards also have provisions for other goals like computational ease and parallelism using tool sets like tiles and slices. Broadly speaking, designers of video coding standards keep the following goals in mind.

- Compression efficiency
- Efficient and parallel implementation especially decoding
- Error resilience
- Transport layer integration

In the following chapters, we will explore in detail the various elements of the codec design that help in achieving each of the above goals.

4.5   SUMMARY

- Video Compression involves an encoder component that converts the input uncompressed video to a compressed stream. This stream, after transmission or storage, is received by a complementary decoder component that converts it back into an uncompressed format.
- Video compression technologies employ frameworks to analyze and compress every picture individually by categorizing them either as intra or inter frame. These identify and eliminate the spatial and temporal redundancies, respectively.
- Inter frames provide significant compression by using unidirectional prediction frames (P frames) and bi-directional prediction frames (B-frames).
- An encoder uses a sequence of periodic and structured organization of I, P and B frames in the encoded stream. This is called a Group of Pictures or GOP.

- I frames are used in the stream to allow fast seek and random access through the sequence.
- Video coding standards use a block-based prediction model in which every frame in turn is split into block-shaped regions of different sizes (e.g. 16x16 macroblocks in H.264, up to 64x64 CTUs in H.265 or 64x64 superblocks in VP9). These blocks are further broken down and partitioned for prediction in a recursive fashion.
- Encoders use slices and tiles to split the frame into multiple processing units so that the encode/decode of these independent units can happen in parallel. These speed up the computations in parallel architectures and multi-threaded environments.

4.6   NOTES

1. Sullivan GJ, Ohm J, Han W, Wiegand T. Overview of the High Efficiency Video Coding (HEVC) standard. *IEEE Trans Circuits Syst Video Technol.* 2012;22(12):1649-1668.
    https://ieeexplore.ieee.org/document/6316136/?part=1. Accessed September 21, 2018.

5   INTRA PREDICTION

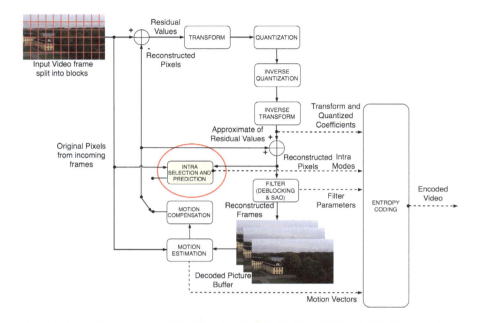

The term, intra frame coding, means that compression operations like prediction and transforms are done using data only within the current frame and not other frames in the video stream.

Every block in an intra frame is thus coded using intra-only blocks. Every frame or block that is intra coded is temporally independent and can be fully decoded without any dependency on other pictures.

Intra-only blocks that use only spatial prediction can also be present in inter frames along with other inter blocks that use temporal prediction. In this chapter we will explore how spatial prediction is performed using different coding tools. Toward the end of the chapter, we will also compare how these tools differ across video standards.

5.1   THE PREDICTION PROCESS

While earlier standards like MPEG2 and MPEG4-Part 2 did not employ spatial pixel intra prediction, this has become a critical step in intra coding in standards since H.264. In this book, we will refer only to spatial pixel intra prediction. This mechanism of prediction exploits the correlation among neighboring pixels by using the *reconstructed* pixels within the frame to derive predicted values through extrapolation from *already coded pixels*. The predicted pixels are then subtracted from the current pixels to get residuals that can be efficiently coded. The objective of intra prediction, therefore, is to find the best set of predicted pixels to minimize the residual information. It uses the pixels from its immediately preceding spatial neighbors and derives the predicted pixels by filling whole blocks with pixels extrapolated from neighboring top rows or left columns of pixels.

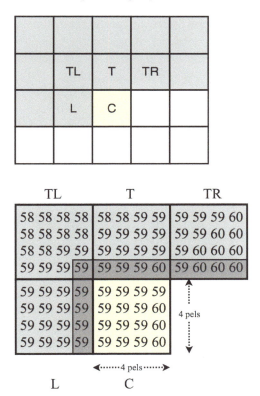

Figure 32: Illustration of neighbor blocks used for intra prediction.

It should be noted that the prediction block is derived using blocks that are previously encoded and reconstructed (before filtering). As blocks in a frame are usually processed in left to right and top to bottom raster fashion, the top and left blocks of the current block are already encoded and hence these available pixels can be leveraged to predict the current block. The left and top pixel sets are double the current block's height and width, respectively, and different codecs limit the usage of pixels from this set for prediction. For example, H.265 uses all the left and bottom left pixels whereas VP9 allows the use of only the left set of pixels.

The concept of using neighboring block pixels for intra prediction is illustrated in Figure 32. The shaded blocks in the figure are all causal. This means that they have already been processed and coded (in scan order) and can be used for prediction. For current block C to be coded, the immediate neighbor blocks are left (L), top (T), top left (TL) and top right (TR). The bottom half of the figure also shows the pixel values for a sample 4x4 luma block and the corresponding luma 4x4 neighboring blocks. The prediction involves the samples from these neighboring blocks that are closest to the current block. The highlighting emphasizes that these samples are very similar and have almost the same values as the luma samples from the current 4x4 block. Intra prediction takes advantage of exactly this pixel redundancy by finding the best of these neighboring pixels that can be used for optimal prediction. This enables use of the fewest bits. Now that we know what intra prediction does, let us explore through an example how it is accomplished.

Example: Let us use the circumstances illustrated in Figure 32 as an example to understand, from an encoder's perspective, which of these pixels best help to predict the current pixels in the given 4x4 block. Let's try the left set of pixels meaning the right-most column of the left neighbor L. In this case, every pixel from this column is duplicated horizontally, as shown in Figure 33(a). Another option is to use the top set of pixels. In this case, this would be the bottom-most row of the top neighbor block T. In this scenario, every pixel from this row is duplicated vertically as shown in Figure 33(b). Projections of pixels along other angles are also possible and an example is shown in Figure 33(c) where the pixels are projected from the bottom-most row of the top (T) and top right neighbor (TR) block, along a 45-degree diagonal.

We have seen in the above example a few possible prediction candidates. Every encoding standard defines its own list of permissible prediction candidates or prediction modes. The challenge for the encoder now is to choose one for every block. The process of intra prediction in the encoder therefore involves iterating through all the possible neighbor blocks and prediction modes allowed by the standard to identify the best prediction mode and pixels for minimizing the number of resulting residual bits that will be encoded.

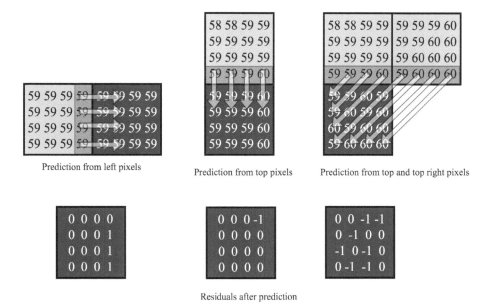

Prediction from left pixels   Prediction from top pixels   Prediction from top and top right pixels

Residuals after prediction

Figure 33: Intra prediction from neighbor blocks by using different directional modes: (a) horizontal, (b) vertical, (c) diagonal.

To do this, the encoder could, for each of the prediction modes, use a distortion criterion like *minimal sum of absolute differences* (min-SAD). This involves a simple computation and indicates the energy contained in the residuals. By computing the SAD for all the modes, the encoder can pick the prediction block that has the least SAD. The SAD, while providing a distortion metric, does not quantify anything about the resulting residual bits if the mode were chosen. To overcome this limitation, modern encoders compute a bit estimate and use it to derive a cost function. This is a combination of the distortion and bits estimate. The best prediction mode is the one that has the most minimal cost function. In the example, as illustrated in Figure 33, the residual blocks contain much smaller numbers. These are cheaper to represent than the original pixels. The SADs of the residual blocks from the three directional modes are 3, 1 and 7, respectively, and the encoder in this case could pick the vertical prediction mode as the best mode if it were using a min-SAD criterion.

The number of such allowable prediction directional modes and block sizes for intra prediction are different across codecs. More modes amount to an increase in complexity in the encoder and better compression efficiency. For example, H.264 allows every 4x4 or 8x8 blocks within a 16x16 macroblock to select a mode from the defined nine intra modes and it also offers one mode for a 16x16 luma block. In VP9, ten prediction modes are defined. These are calculated for the 4x4 block.

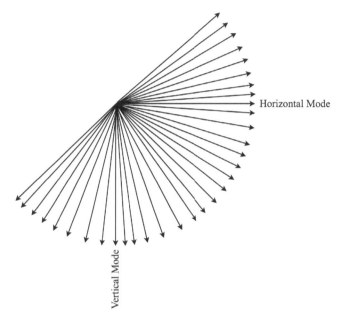

Figure 34: Intra prediction angular modes defined in H.265.

H.265, on the other hand, extends it further to use 35 modes for prediction of 4x4, 8x8, 16x16 or 32x32 sub blocks within the CTU. The prediction direction of the 35 modes is shown in Figure 34. The increased angles in H.265 are designed such that they provide more options for near-horizontal and near-vertical angles and fewer options for near-diagonal angles, in accord with statistical observations.

The upcoming AV1 codec has up to 65 different modes. It's imperative, therefore, that H.265 and AV1 have better prediction and thereby video quality improvement. However, they also have significant computational complexity increase relative to other standards like H.264.

The most common prediction modes available across all codecs are directional. This includes horizontal, vertical and other angular directions, a DC predictor and any other standard-specific specialized modes (e.g., VP9 has a specialized mode called True Motion, or TM mode). Typically, the number of directional modes varies widely across standards, as we have discussed above.

While the exact number of pixels and the number of prediction modes differ slightly across codecs, conceptually they are all the same, as explained above.

When the intra prediction modes are established, the following are then sent as part of the encoded bit stream for every intra predicted block:

a)   The prediction mode
b)   Residual values (differentials of current and predicted pixels)

The decoder performs a complementary operation when it receives the bitstream. It uses the intra prediction mode to look at the already decoded pixels and form the predicted pixels. It then adds this to the residuals from the bitstream to arrive at the pixels of the current block.

5.2   TRANSFORM BLOCKS AND INTRA PREDICTION

Intra prediction for a block that's encoded in intra mode is done by successively doing prediction on all the smaller blocks of pixels within the block. For example, doing intra prediction on each of the sixty-four 8x8 blocks within a superblock.

The prediction process is similar in H.264/H.265 and VP9 and is done for every transform block. This means that, if the transform block size is 8x8 and the block partition is 16x16, there will be one prediction mode for the entire partition but intra prediction will be done for every 8x8 block.

However, if the transform block size is 32x32 and the partition size is 16x16 (this is permitted in H.265), there will be only one intra prediction mode shared by the entire 32x32 block.

It should be noted that, as transform sizes are square, intra prediction operations are always square.

Figure 35: Original raw source.

Some standards, like VP9, allow separate intra prediction for luma and chroma in which case the process can be repeated for chroma. H.264 and H.265, however, use the same luma intra prediction mode for chroma. The figures 35-37, illustrate the process and efficiency of spatial intra prediction for a complete intra frame. Figure 35 shows the raw source frame.

Figure 36: Image formed from intra predicted pixels.

Figure 37: Residual image formed by subtracting original and predicted pixel values.

Figure 36 shows the image formed by the predicted pixels. Notice how close the predicted pixels are to the original source in Figure 35. Finally, Figure 37 shows the residual values and by looking at this image, we can infer the accuracy of the prediction. It can be observed that the prediction is quite accurate even at a block level in flat areas like the sky, whereas in places of detail like the building windows and the trees, the blocks used in prediction are unable to completely capture these details. This results in prediction errors that will eventually be encoded in the bitstream as residual values.

5.3   COMPARISON ACROSS CODECS

As highlighted earlier, different video coding standards have similar intra prediction principles but differ in the following fine details:

- Transform Sizes
- Number and type of intra prediction modes
- Variations in handling luma and chroma pixels

Table 7 highlights the differences in the above characteristics that define the intra prediction process in a few video standards. We notice that more angles are used in newer codecs to derive increased prediction efficiency at the expense of increased computational complexity.

Table 7: Comparison of intra prediction across codecs.

Feature	H.264	H.265	VP9
Transforms	4x4 integer DCT transforms	variable 32x32 down to 4x4 integer DCT transforms + 4x4 integer DST transform	variable 32x32 down to 4x4 integer DCT transforms + 4x4 integer DST transform
Intra prediction	spatial pixel prediction with 9 directions	spatial pixel prediction with 35 directions	8 angles for spatial directional prediction.

5.4   SUMMARY

- The term, intra frame coding, implies that the compression operations like prediction and transforms are all done using data only within the current frame and no other frames in the video stream.
- Intra prediction exploits the correlation among neighboring pixels by using the reconstructed pixels within the frame to derive predicted values through extrapolation from already coded pixels.
- Reconstructed pixels from the last columns and rows of the left and top neighboring blocks, respectively, are typically used for prediction of the pixels of the current block.
- Encoders choose the pixels to predict from by selecting from several directional modes and block sizes. More modes result in an increase in complexity in the encoder but provide better compression efficiency.
- The prediction mode and residual values are signaled in the encoded bit stream for every intra predicted block.

- One intra prediction mode is selected for every partition but the prediction process is done for every transform block. For example, if transform size is 8x8 and the partition is 16x16, there will be one prediction mode for the 16x16 partition but intra prediction will be performed for every 8x8 block.

6 INTER PREDICTION

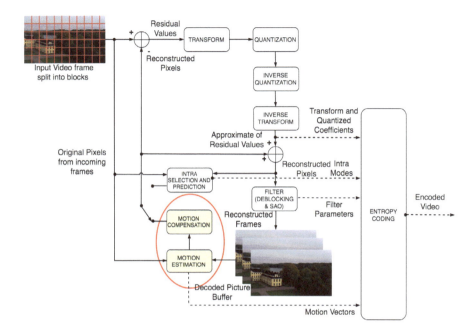

Inter frame coding implies that the compression operations like prediction and transforms are done using data from the current frame and its neighboring frames in the video stream. Every frame or block that is inter coded is dependent on temporally neighboring frames (called reference frames). It can be fully decoded only after the frames on which it depends are decoded.

Unlike in intra prediction, where pre-defined modes in the standard define from which directions blocks may be used for prediction, inter prediction has no defined modes. Thus, the encoder has the flexibility to search a wide area in one or several reference frames to derive a best prediction match. The partition sizes and shapes within a superblock are also flexible such that every sub-partition could have its best matching prediction block spanning different reference frames.

In intra prediction, a row or column of pixels from the neighboring blocks are duplicated to form the prediction block. In inter prediction, there is no extrapolation mechanism. Instead, the entire block of pixels from the reference frame that corresponds to the best match forms the prediction block.

6.1   MOTION-BASED PREDICTION

As different objects in the scene can move at different speeds, independently of the frame rate, their actual motion displacements are not necessarily in integer pel units. This means that limiting the search for a block match in the reference frames using pixel granularity can lead to imperfect prediction results. Searching using sub-pixel granularity could give better matching prediction blocks, thereby improving compression efficiency. The natural question, then, is how to derive these sub-pixels, given that they don't actually exist in the reference frames? The encoder will have to use a smart interpolation algorithm to derive these sub-pel values from the neighboring full-pel integer samples. The details of this interpolation process are presented in later sections of this chapter. For now, we assume that the encoder uses such an algorithm and searches for matching prediction blocks with sub-pixel accuracy by interpolating pixel values between the corresponding integer-pixels in the reference frame. If it still turns out that good temporal matching blocks are unavailable or intra prediction yields better results, the block is coded as intra.

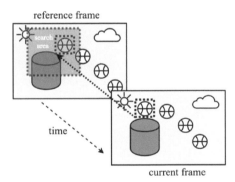

Figure 38: Deriving the motion vector using motion search.

The process of searching the reference frames to come up with the best matching prediction block is called *motion estimation* (ME). The spatial displacement of the current block from its prediction block is called *motion vector* (MV) and is expressed in (X, Y) pixel coordinates. Figure 38 illustrates this concept. The motion search for the sample block that contains a ball is shown. In the reference frame, among the other blocks in the defined search area, the highlighted block containing the ball is chosen and the relative distance between the current and the chosen block is tracked as the motion vector. The mechanism to use motion vectors to form the prediction block is called *motion compensation* (MC). It is one of the most computationally intensive blocks in an encoder.

The prediction block derived using motion search is not always identical to the current block. The encoder, therefore, calculates the residual difference by subtracting the current block pixel values from those of the prediction block. This, along with the motion vector information, is then encoded in the bitstream.

The idea is that a better search algorithm results in a residual block with minimal bits, resulting in better compression efficiency. If the ME algorithm can't find a good match, the residual error will be significant. In this case, other possible options evaluated by the encoder could include intra prediction of the block or even encoding the raw pixels of the block.

Also, as inter prediction relies on reconstructed pixels from reference frames, encoding consecutive frames using reference frames that have previously been encoded using inter prediction often results in residual error propagation and gradual reduction in quality. An intra frame is inserted at intervals in the bitstream. These reset the inter prediction process to gracefully maintain video quality.

Once the motion vector is derived, it needs to be signaled in the bitstream and this process of encoding a motion vector for each partition block can take a significant number of bits. As motion vectors for neighboring blocks are often correlated, the motion vector for the current block can be predicted from the MVs of nearby, previously coded blocks. Thus, using another algorithm, which is often a part of the standard specification, the differential MVs are computed and signaled in the bitstream along with every block.

When the decoder receives the bitstream, it can then use the differential motion vector and the neighboring MV predictors to calculate the absolute value of the MV for the block. Using this MV, it can build the prediction block and then add to it the residuals from the bitstream to recreate the pixels of the block.

Figure 39: Frame 56 of input - stockholm 720p YUV sequence.

Figure 40: Motion compensated prediction frame.

The figures 39-42 illustrate step-by-step how a video frame is encoded and decoded using block-based inter prediction. Figure 39 shows the input clip.

Its motion compensated prediction frame is shown in Figure 40 alongside the corresponding motion vectors in Figure 41.

Figure 41: Motion vectors from reference frames.

Figure 42: Motion compensated residual frame [1].

Figure 42 shows the residual frame that is obtained as a difference between the original frame and its prediction. We notice from Figure 40 how visually close the prediction frame is to the actual frame that is encoded. This is also objectively represented in Figure 42. It is mostly gray, representing areas of strong similarity between the predicted and actual pixels.

6.1.1   MOTION COMPENSATED PREDICTION

As we've seen in earlier sections, shifted areas in the reference frames are used for prediction of blocks in the current frame. Using the process of motion estimation, a displacement motion vector is derived. It corresponds to the motion shift between the reference and current frame for the block in question.

6.1.1.1   BIDIRECTIONAL PREDICTION

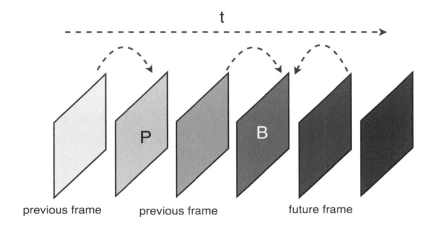

Figure 43: Illustration of bidirectional prediction.

Motion compensated prediction can be performed for every block using either one frame or two frames as reference. Thus, a block could potentially have multiple sources based on which it can predict, and this number depends on whether the current block belongs to a P frame or a B frame. As shown in Figure 43, in the case of P frames, only one prediction reference frame is allowed for every block whereas in the case of B frames, every block could predict up to two reference frames. Encoders like H.264 and H.265 maintain two separate reference frame lists (List L0 and List L1) where any block of a P frame can predict using reference frames from List L0. A block from B frame can predict using either L0 or L1 or both L0 and L1, where one reference frame from each L0 and L1 can be used.

In using reference lists, it's entirely up to the encoder implementation how these lists are managed, including how many pictures are added in these lists and what pictures get added to each list (as long as they are within the profile/limits set by the standard). The same picture can be added to both lists and, interestingly, this could be used to simulate motion vectors with higher ⅛ pixel precision (without actual bitstream signaling). This can be done for any block by using two motion vectors that are a quarter pixel apart and point to the same reference picture.

6.1.1.2   WEIGHTED PREDICTION

Block-based inter prediction is sensitive to illumination variations between frames, especially quick changes in illumination caused by flashes fired during live events or fades-ins and fade-outs. These can cause significant variation in intensity across immediately successive frames, which may result in poor motion estimation and compensation.

Fade Out

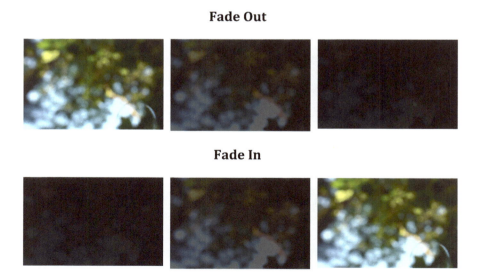

Fade In

Figure 44: Fades in video sequences.

Fades are illustrated in Figure 44. We can observe that the intensity varies consistently but dramatically across successive frames.

Weighted prediction (WP) is a tool available in H.264 and H.265. It helps to overcome the challenges due to quick illumination changes by applying a weighting factor and offset during the process of prediction from the reference frame. By using WP, the pixels in the reference frame are first scaled using a multiplication factor, W, and then shifted by an additive offset, O. During the process of motion compensation, the Sum of Absolute Differences (SAD) between the current frame f_{curr} and the reference frame f_{ref} can be mathematically expressed as:

$$SAD_{WP} = \sum_m | f_{curr} - F_{ref}| \text{ where } F_{ref} = W * f_{ref} + O$$

The challenge here is to derive the correct values of W and O, however, when found, the SAD can be correspondingly compensated for and the ME process can accurately get the correct matching prediction block. It should be noted that different situations would warrant slightly different approaches. Scenes with fade-in and fade-out have global brightness variations across the whole pictures. This can be compensated for correspondingly with frame level WP parameters. However, scenes with camera flash will have local illumination variations within the picture which will obviously require more localized WP parameters for efficient compensation. This is also true in scenes with fade-in from white where the brightness variation of blocks with lighter pixels is smaller than that of darker pixels. Localized WP parameters, however, will introduce excessive overhead bits in the encoded bitstream and are not available in H.264 and H.265. To combat this, various approaches have been developed that effectively use the available structures of multiple reference frames with different WP parameters to compensate for non-uniform brightness variations.

6.2   MOTION ESTIMATION ALGORITHMS

In differential coding, the prediction error and the number of bits used for encoding depend on how effectively the current frame can be predicted from the previous frame. As we discussed in the earlier sections, if there's motion involved, the prediction for the moving parts in the sequence involves motion estimation (ME); that is, finding out the position in the reference frame from where it has been moved.

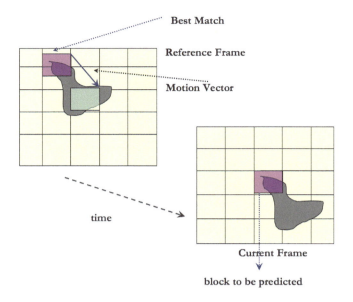

Figure 45: Block-based motion estimation.

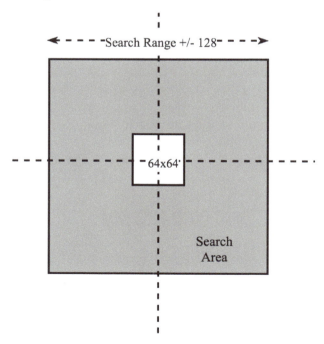

Figure 46: Search area with range +/- 128 pixels around a 64x64 block.

Figure 45 illustrates the block-based ME process. In this process, the best match is the block that gives the minimum sum of absolute error between the two blocks. As various objects in a frame could move to any position in the next frame, depending on their velocities, searching the entire set of reference frames for the suitable predictor would be very computationally intensive. Typically, between successive frames, the motion of a particular object would be restricted to a few pixels in both the vertical and horizontal directions.

Hence, motion estimation algorithms define a rectangular area enclosing the current block called the *search area* to conduct the search. The search area is usually specified in terms of a *search range.* This gives the horizontal and vertical number of pixels to search for the predictor block. Figure 46 shows a typical search region around a 64x64 block with a search range of +/-128 pixels.

It is expected that the best motion search match is to be found in one of the points within this search range. However, it should be noted that this is not guaranteed. Sometimes the motion can be quite considerable and outside this range, as well; for example, a fast-moving sports scene. In such cases, the best match could be the motion point in the search area that is closest to the actual motion vector. Intra prediction can also be used if there's no suitable motion vector match. Depending on how and what motion points are searched in the motion estimation process, the search algorithms can be classified as follows: *Full search* or *exhaustive search* algorithms employ a brute force approach in which all the points in the entire search range are visited. The SAD or similar metrics at all these points are calculated and the one with the minimum SAD is adjudged the predictor. This is the best of all the search techniques, as every point in the search range is evaluated meticulously. The downside of exhaustive search is its excessive computational complexity.

Smart search algorithms try to overcome the heavy computational load imposed by exhaustive search algorithms by not evaluating all the candidate points. Instead, they use specific search patterns to dramatically reduce the number of search points to find the motion vector. Many fast search algorithms have been developed that differ primarily in the pattern in which the search points are evaluated. Examples are 2-D logarithmic search (LOGS), three-step search (TSS), diamond search (DS), and so on. In this section, we shall illustrate a simple 3-Step logarithmic search algorithm. It is illustrated in Figure 47, where a sample search area of +/- 7 pixels is used.

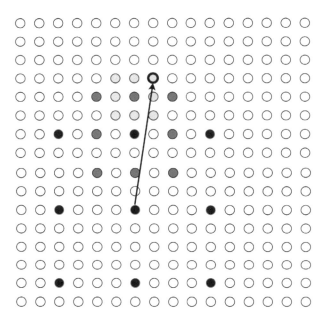

Figure 47: Three step search for motion estimation.

The search in this algorithm is performed in three separate steps and each step progresses from a coarse-grained area to a finer-grained search area. Every dot here represents a search point. The black dots represent the points visited in the first step. These are widely spaced at a distance of 4 pixels apart. Nine such points at square locations are searched and the one with the minimum cost is selected as the best point of this step. The search continues by keeping this best point identified in step 1 as the center and evaluating the 8 square points around it. These are only 2 pixels distance apart, as shown by the dark gray dots. The best point in this second step is then used as the center for the final search of 8 square points around it. These points are only 1 pixel away from each other. The best point of the third step is then chosen as the final integer pixel MV.

This method provides good speedup by using only 25 search points, compared to 196 search points that would be used by exhaustive search for the same search area. As the search range becomes larger, larger square patterns can be used and this mechanism provides higher speedup ratios.

6.3   SUB PIXEL INTERPOLATION

As explained earlier, sub-pixel precision in motion vectors is needed as different objects in the video scene can have motion at speeds that are independent of the frame rate. This cannot be accurately captured using full pixel motion vectors alone. In this section, let us see how such sub-pixel accurate motion vectors are calculated.

Luma and chroma pixels are not sampled at sub-pixel positions. Thus, pixels at these precisions don't exist in the reference picture. Block matching algorithms therefore have to create them using interpolation from the nearest integer pixels and the accuracy of interpolation depends on the number of integer pixels and the filter weights that are used in the interpolation process.

Sub-pixel motion estimation and compensation is found to provide significantly better compression performance than integer-pixel compensation and ¼-pixel is better than ½-pixel accuracy. While sub-pixel MVs require more bits to encode compared to integer-pixel MVs, this cost is usually offset by more accurate prediction and, hence, fewer residual bits.

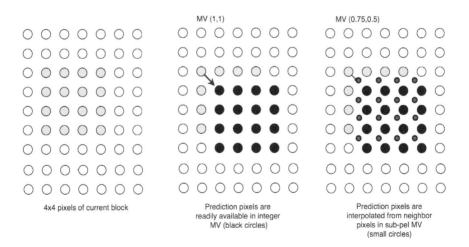

Figure 48: Example of integer and sub-pixel prediction.

Figure 48 illustrates how a 4x4 block could be predicted from the reference frame in two scenarios: integer-pixel accurate and fractional-pixel accurate MVs. In Figure 48a, the grey dots represent the current 4x4 block. When the MV is integral (1,1) as shown in Figure 48b, it points to the pixels corresponding to the black dots that are readily available in the reference frame.

Hence, no interpolation computations are needed in this case. When the MV is fractional (0.75, 0.5), as shown in Figure 48c, it has to point to pixel locations as represented by the smaller gray dots. Unfortunately, these values are not part of the reference frame and have to be computed using interpolation from the neighboring pixels.

H.265 uses the same MVs for luma and chroma and uses ¼ accurate MVs for luma that are computed using six-tap interpolation filters. For YUV 4:2:0, these MVs are scaled accordingly for chroma as ⅛ pixel accurate values. VP9 uses a similar interpolation but uses a longer eight-tap filter and also a more accurate ⅛-pixel interpolation mode.

In VP9, the luma half-pixel samples are generated first and are interpolated from neighboring integer-pixel samples using an eight-tap weighting filter. This means that each half-pixel sample is a weighted sum of the 8 neighboring integer pixels used by the filter.

Once half-pixel interpolation is complete, quarter-pixel interpolation is performed using both half and full-pixel samples.

6.3.1   SUB-PIXEL INTERPOLATION IN HEVC

In this section, the fractional interpolation process is illustrated in detail using the HEVC interpolation filters as an example. The integer and fractional pixel positions used in HEVC, as adapted from Sullivan, Ohm, Han, & Wiegand, [1] are illustrated in Figure 49. The positions labeled $A_{i,j}$ represent the luma integer pixel positions, whereas $a_{i,j}$, $b_{i,j}$ and so on are the ½-pixel and ¼-pixel positions that will be derived by interpolation.

In HEVC, the half-pixel values are computed using an eight-tap filter. However, the quarter-pixel values are computed using a seven-tap filter. Let us now illustrate how all the 15 positions marked $a_{0,0}$ to $r_{0,0}$ are computed. The filter coefficient values are given in Table 8, below. An example for sample $b_{0,j}$ in half sample position and $a_{0,j}$ in quarter sample position is given below.

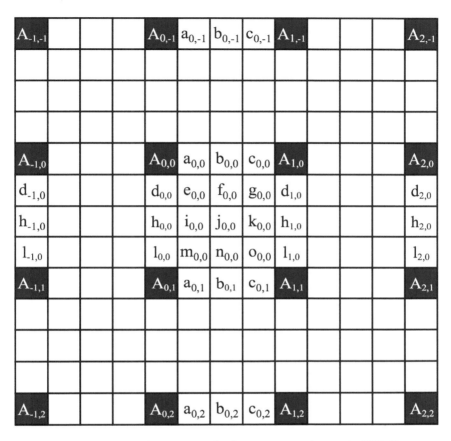

Figure 49: Pixel positions for luma Interpolation in HEVC.

$$a_{0,j} = \sum_{i=-3...3} A_{i,j} \cdot QF[i]$$

$$b_{0,j} = \sum_{i=-3...4} A_{i,j} \cdot HF[i]$$

Table 8: Interpolation filter coefficients used in HEVC.

Index	-3	-2	-1	0	1	2	3	4

HF	-1	4	-11	40	40	-11	4	1
QF	-1	4	-10	58	17	-5	1	

$$e_{0,0} = \left(\sum_{i=-3\ldots3} a_{0,i} \cdot QF[i] \right) >> 6$$

$$i_{0,0} = \left(\sum_{i=-3\ldots4} a_{0,i} \cdot HF[i] \right) >> 6$$

The samples labeled $e_{0,0}$ to $o_{0,0}$ can then be derived by applying the same filters to the above computed samples as shown in the equations above.

At this point, weighted prediction can also be applied, if enabled. The prediction values computed above are scaled and offset using the WP weight and offset that are signaled in the encoder. The fractional sample interpolation process for the chroma components is similar to that for luma. However, a different interpolation filter with only 4 filter taps is used.

Table 9: Chroma interpolation filter coefficients used in HEVC.

Index	-1	0	1	2
1/8	-2	58	10	-2
2/8	-4	54	16	-2
3/8	-6	46	28	-4
4/8	-4	36	36	-4
5/8	-4	28	46	-6
6/8	-2	16	54	-4
7/8	-2	10	58	-2

It should be noted that the fractional accuracy is ⅛-pixel for subsampled chroma in the 4:2:0 case. The four-tap filters for eighth-sample positions (1/8th, 2/8th, 3/8th upto 7/8th) for 4:2:0 chroma are as given in Table 9, above.

6.4   MOTION VECTORS PREDICTION

Encoding absolute values of motion vectors for each partition can consume significant bits. The smaller the partitions chosen, the greater is this overhead. The overhead can also be significant in low bit rate scenarios. Fortunately, as highlighted in Figure 50, the motion vectors of neighboring blocks are usually similar. This correlation can be leveraged to reduce bits by signaling only the differential motion vectors obtained by subtracting the motion vector of the block from the best neighbor motion vector. A predicted vector, PMV, is first formed from the neighboring motion vectors. DMV, the difference between the current MV and the predicted MV, is then encoded in the bitstream.

Figure 50: Motion vectors of neighboring blocks are highly correlated [2].

The question now is, which neighboring MV is most suitable for prediction of any block? Different standards allow different mechanisms to derive the PMV. It usually depends on the block partition size and on the availability of neighboring MVs. Both HEVC and VP9 have an enhanced motion vector prediction approach. That is, MVs of several spatially and temporally neighboring blocks that have been coded earlier are candidates that are evaluated for selection as the best PMV candidate. In VP9, up to 8 motion vectors from both spatial and temporal neighbors are searched to arrive at 2 candidates. The first candidate uses spatial neighbors, whereas the second candidate list consists of temporal neighbors.

VP9 specifically prefers to use candidates using the same reference picture and searches this picture first. However, candidates from different references are also evaluated if the earlier search fails to yield enough candidates. If there still aren't enough predictor MVs, then 0,0 vectors are inferred and used. Once these motion vector predictors (PMVs) are obtained, they are used to signal the DMV in the bitstream using either of the four modes available in VP9.

Three out of the four modes correspond to direct or merge modes. In these modes, no motion vector differential need be sent in the bitstream. Based on the signaled inter mode, the decoder just infers the predictor MV and uses it as the block's MV. These modes are as follows.

- *Nearest MV* uses the first predictor intact, with no delta.
- *Near MV* uses the second predictor with no delta.
- *Zero MV* uses 0,0 as the MV.

In the fourth mode, called the New MV mode, the DMVs are explicitly sent in the bitstream. The decoder reads this motion vector difference and adds it to the nearest motion vector to compute the actual motion vector.

- *New MV* uses the first predictor of the prediction list and adds a delta MV to it to derive the final MV. The delta MV is encoded in the bitstream.

H.265 also uses similar mechanisms as above with slightly different terminologies and candidate selection process. In H.265, the following modes are allowed for any CTU.

Merge Mode. This is similar to the first three modes of PMV in VP9 where no DMV is sent in the bitstream and the decoder infers the motion information for the block using the set of PMV candidates. The algorithm on how to arrive at the specific PMV for every block is specified in the standard.

Advanced Motion Vector Prediction. Unlike the merge mode, in this mode the DMV is also explicitly signaled in the bitstream. This is then added to the PMV (derived using a process similar to the above for merge mode) to derive the MV for the block.

Skip Mode. This is a unique mode that is used when there is motion of objects without any significant change in illumination. While earlier standards defined skip mode to be used in a perfectly static scenario with zero motion, newer codecs like H.265 defined it to include motion. In H.265, a skip mode syntax flag is signaled in the bitstream and, if enabled, the decoder uses the corresponding PMV candidate as the motion vector and the corresponding pixels in the reference frame as is, without adding any residuals.

6.5   SUMMARY

- In inter frame coding, operations like prediction and transforms are done using data from the current and neighboring frames in the video stream.
- The process of searching the reference frames to come up with the best matching prediction block is called motion estimation (ME) and the mechanism to use the motion vectors (MVs) to form the prediction block is called motion compensation (MC).
- Sub-pixel precision in motion vectors is needed as different objects in the video scene can have motion at speeds that are independent of the frame rate and hence cannot be represented using full pixel MVs.
- To overcome motion vector signaling overhead, MVs of neighboring blocks, which are quite similar, are used to predict block MV and only resulting residual MV is signaled in the bitstream.

6.6   NOTES

1.   Sullivan GJ, Ohm J, Han W, Wiegand T. Overview of the High Efficiency Video Coding (HEVC) standard. *IEEE Trans Circuits Syst Video*

Technol. 2012;22(12):1649-1668.
https://ieeexplore.ieee.org/document/6316136/?part=1. Accessed
September 21, 2018.

2. VP9 Analyzer. Two Orioles. https://www.twoorioles.com/vp9-
   analyzer/. Accessed September 22, 2018.

7 RESIDUAL CODING

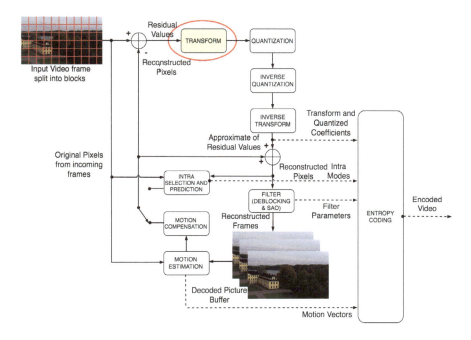

After prediction is completed, the block of residual values undergoes a series of processes before they become bits and bytes in the encoded bitstreams. These processes include transforms and quantization stages, and these are covered in detail in this chapter. The transform stage takes in the block of residual values after prediction and converts it to a different domain called frequency domain. It's the same set of values but represented differently in the frequency domain.

7.1 WHAT ARE FREQUENCIES?

Simply put, frequency refers to the rate at which something is repeated over a particular period of time. The more it's repeated, the higher is the frequency and *vice versa*. Frequency is thus the inverse of the time period of the change. This means that the shorter the time it takes for the change from one value to another and back, the higher is the frequency of occurrence of that value over the time period. In the case of pictures, pixel values vary in intensities and the time it takes to *change* from one intensity to another and back again is represented by frequency. The faster the change of intensity from, say, light to dark and back, the higher the frequency needed to represent that part of the picture.

In other words, frequency in a picture is nothing but a representation of the rate of change. Rapidly changing parts of the picture (e.g., edges) contain high frequencies and slowly changing parts (e.g., solid colors and backgrounds) contain low frequencies.

Let us look at an example. Say the block of image pixels is black. It doesn't display any *change*, meaning it has an infinite interval of change and thus low frequency. Now, however, if the block of image pixels is black at the left, turns white in the center and then turns black again, it then has one change in the interval until it's back to its original value. This means it has a finite frequency of, say, one. If the block has two such changes, then it has a higher frequency of two and so on.

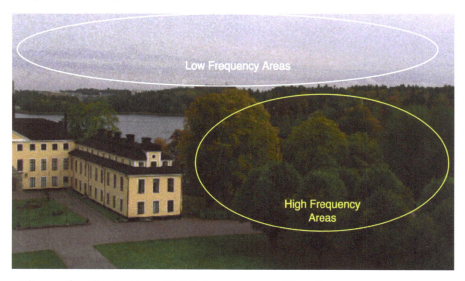

Figure 51: Illustration of high frequency and low frequency areas in an image.

Thus, the greater the number of changes in the values in the period in question, the greater is the frequency. This is illustrated in Figure 51 for an image from the into tree clip. The sky area has low variations in pixel values and, correspondingly, low frequency components. In contrast, the trees have significant texture, hence higher variations in pixel intensities and high frequency components.

7.2   HOW CAN AN IMAGE BE BROKEN DOWN INTO ITS FREQUENCIES?

Within every block of pixels in the image, we can approximate the individual column and row of pixels as the sum of a series of frequencies, starting with the lowest frequency and adding more frequencies. The block of pixels is thus a juxtaposition of a series of frequencies. The lowest frequency, in effect the DC or average value of pixels in the block, doesn't add any *fine details* at all. With every frequency added, one after another, more and more details are built up in the picture.

The basic idea here is that a complex signal like an image pixel block can be broken down into a linear weighted sum of its constituent frequency components. Higher frequency components represent more details.

7.2.1   WHY MOVE TO THE FREQUENCY DOMAIN?

The pixel blocks in the image have components of varying frequencies and the transform process serves to represent it as a linear weighted sum of its constituent frequency components. The sections of high detail like edges correspond to high frequency components. The flat areas correspond to low frequency components, as described in the previous section. Splitting the image in this way affords us certain advantages, as follows.

7.2.1.1   ENERGY COMPACTION

Blocks of video picture samples exhibit strong spatial correlation. This also extends to residuals, as illustrated in Figure 52 for a sample 32x32 block. We see here the residuals are not only similar, but also smaller values that can therefore be efficiently represented. Similarities and correlation are more synonymous to flat transitions as opposed to dramatic changes. What this means is the energy or frequency in the pixels is usually concentrated around the low frequency components relative to the high frequency components. This concentration of energy is called *energy compaction* and it is a critical reason why transforms are needed.

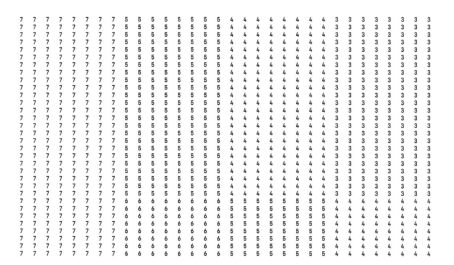

Figure 52: Residual sample values for a 32x32 block.

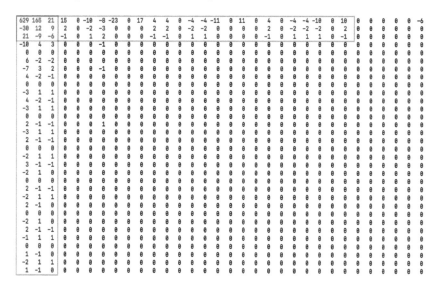

Figure 53: Energy compaction after transforms.

With energy compaction, we are able to group these components in the order of increasing frequencies and obtain a pattern wherein the higher energy, low frequency components occur at the start and gradually taper to the lower energy, high frequency components. This allows us a very efficient representation of the block of samples with much lower values.

Figure 53 shows the 32x32 block of Figure 52 after it is transformed using a *discrete cosine transform* that is described in a later section. From this figure, we clearly see how the transformed 32x32 block exhibits energy concentration toward the high frequency components; that is, around the top left of the block after the transformation.

As we move away from the top left area, we see a lot of small values and redundant zeros that can be efficiently represented after a suitable reordering process. The details of how reordering is done are covered in subsequent sections. For now, it suffices to note that the reordering process helps in efficiently representing the transformed values by making use of the energy patterns.

7.2.1.2 DISCARDING TOWARD COMPRESSION

Once we do the transform that results in energy compaction, we have the ability to analyze this data set and go one step further to selectively discard some of the data. As the HVS is more sensitive to uniform areas than to more detailed areas, we could focus on low frequency components and selectively discard the highest frequency components. Though discarding the high frequency components means losing some of the detail, this process is found to produce a good approximation of the original block at the expense of some level of smoothing.

The amount of data kept and discarded depends on the appetite for detail needed and the level of compression required. Furthermore, the HVS is more sensitive to luma information than chroma. This means we can be slightly more aggressive with this process for chroma than for luma. The process of discarding the data is called *quantization* and is covered in detail in upcoming sections.

7.2.2   CRITERIA FOR TRANSFORM SELECTION

When selecting transforms, the following are key criteria.

7.2.2.1 ENERGY COMPACTION

The transform should provide a complete decorrelation of the data and maximum energy compaction. We have explained this in the previous section.

7.2.2.2 THE TRANSFORM SHOULD BE REVERSIBLE

This means when the decoder receives these transformed coefficients, it should be able to do an inverse transform operation to retrieve the input samples accurately. It should be noted that this process is also used by the encoder as it performs this sub-section of the decoder tasks to store the reconstructed pixels internally.

7.2.2.3 THE TRANSFORM SHOULD BE IMPLEMENTATION FRIENDLY

This is truer of the process of inverse transforms that are implemented by the decoder. As a variety of decoding platforms with a host of capabilities exist, it's useful for the video to be decodable by a maximum number of decoders with a minimal complexity requirement. Common requirements include minimal memory storage requirement, less arithmetic precision for storage of internal computation results and fewer arithmetic operations.

7.2.3   DISCRETE COSINE TRANSFORM

Now that we have seen what transforms are, why they are needed, and the criteria used for their selection, let us explore some popular ones. Over the years, several transforms have been proposed and used in different video and image coding standards, including the popular *discrete cosine transform* (DCT), *discrete sine transform* (DST) and *Hadamard transform*.

These transforms operate on a block of image or residual values and fit very well in the block-based video compression framework. In this section, we will focus on the landmark DCT that has been widely employed since the era of the MPEG2 video standard, thanks to its simplicity and high energy compaction. The concepts explained are fundamental and similar across the various transforms.

DCT expresses the input signal as a sum of sinusoids of different frequencies and amplitudes. It is similar to the discrete Fourier transform but uses only cosine functions and real numbers.

In most video standards, a block of residual values is transformed using a 4x4 or 8x8 integer transform that is an approximate of the DCT and it will be the focus of this book.

$$F(u,v) = \frac{2c(u).c(v)}{N} \sum_{m=0}^{N-1} \sum_{n=0}^{N-1} f(m,n).cos(\frac{2m+1}{2N}u\pi)cos(\frac{2n+1}{2N}v\pi)$$

$$where \; u,v = 0,1,...,N-1$$

$$c(k) = \frac{1}{\sqrt{2}}k = 0$$

$$= 1 \quad otherwise$$

The above is the equation of the two-dimensional DCT. While it looks complicated, the equation can be understood easily as follows: Any two-dimensional signal f can also be expressed as F, its transformed counterpart and F is the weighted sum of component values of f with cosine basis functions. The equation is two-dimensional, as it uses two-dimensional cosine basis functions corresponding to u and v. The value N is chosen corresponding to the size of the residual matrix. In the above equation, when N = 4 it yields the 4x4 matrix of 2D cosine transform coefficients as:

0.5	0.5	0.5	0.5
0.653	0.271	0.271	-0.653
0.5	-0.5	-0.5	0.5
0.271	-0.653	-0.653	0.271

4x4 DCT Coefficients A =

The DCT of a set of 4x4 samples **X** is given by the expression below, where Y is the transformed block, **A** is the DCT basis set matrix, and **A**^{Tr} is its matrix transpose.

$$\mathbf{Y = A \, X \, A^{Tr}}$$

Figure 54 shows a set of 8x8 residual values that is taken from the top left, 8x8 block of a sample 16x16 block of residual values. Performing the 8x8 DCT operation yields the 8x8 matrix shown in Figure 55. It should be noted from Figure 55 that the larger coefficients are compactly located around the top left corner, in other words, around the low frequency DC component. This is the desired energy compaction function of the transform operation. H.265 and VP9 define prediction modes in accordance with the transform sizes and also use a combination of a few transforms to suit different prediction modes. VP9 supports transform sizes up to the prediction block size, or, 32x32, whichever is less.

```
-1    -1    -1    -1    -1    -1    -1    -1
 0     0    -1    -1    -1    -1    -1    -1
 1     0    -1    -1    -2    -2    -2    -2
 1     1     0    -1    -2    -2    -3    -3
 2     2     1     1     0    -1    -2    -2
 2     2     1     1     0     0    -1    -2
 0     0     1     1     0     0    -1    -1
-3    -2     0     0     1     1     0     0
```

Figure 54: Residual samples: top-left 8x8 block.

```
-3.875   5.065  -0.856   0.176  -0.125  -0.084   0.219  -0.606
-3.272   0.906   2.531   0.370   0.301   0.025  -0.573  -0.084
-0.490  -5.969  -1.369  -0.339  -0.163  -0.030   0.390   0.587
 2.535   1.594  -0.214   0.199  -0.003   0.098  -0.340  -0.236
-0.875  -0.909  -0.856  -0.357  -0.125  -0.513   0.219   0.307
-1.334   0.840   0.250  -0.329  -0.127   0.374   0.396  -0.113
-0.203  -0.267   0.140   0.410  -0.068   0.221  -0.131   0.016
 0.387  -0.011  -0.106  -0.318   0.675   0.267   0.024   0.021
```

Figure 55: 8x8 DCT coefficients of the residual samples.

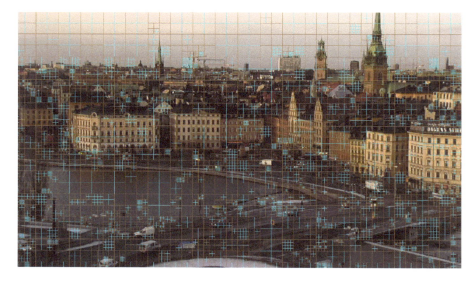

Figure 56: Flexible use of different transform sizes.

Figure 56 shows a screenshot from the *stockholm* clip with the grids showing the transform sizes. Larger transform sizes, up to 32x32, are used in the smooth areas like the sky and water, while smaller transform sizes are better able to capture the fine details like buildings and so on. Encoders, in addition to deciding the best prediction modes for every block, also have to decide the optimal transform size for every block.

7.3   QUANTIZATION

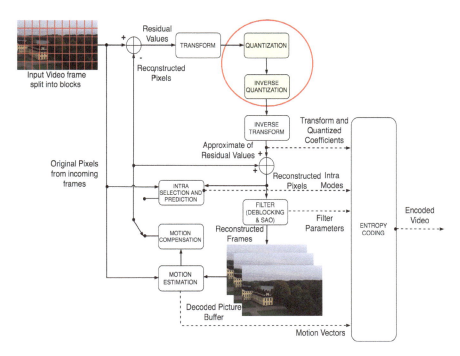

7.3.1   THE BASIC CONCEPTS

Quantization is a process of reducing the range of a set of numbers. The range is an important consideration as it determines the number of values and hence the number of bits needed to represent every value. With a reduced range, the numbers in the new number set can thus be represented using fewer bits. This also means that the granularity of values in the resulting number set is also reduced. In general, the higher the quantization that is applied, the coarser is the resulting data set. Quantization can be achieved by a simple process of division at the encoding side and a corresponding multiplication at the reception or decoding side. Let's illustrate how this is done using an example. Let's consider a set of integers with a range of values from 0 to 255, as shown in the original 4x4 matrix in Figure 57a, below. When the numbers in this set are divided by a fixed value, say, 4, the resulting numbers will only have a range from 0 to 63, meaning only 64 values. Dividing by 4 and discarding the remainders, assuming this is an integer set, we get the following quantized matrix in Figure 57b. When we do the reverse operation of multiplication (by the same value 4) at the receiving end, we obtain the reconstructed 4x4 matrix as shown in Figure 57c below.

240	123	42	4		60	30	10	1		240	120	40	4
239	51	12	3		59	12	3	0		236	48	12	0
81	13	3	0		20	3	0	0		80	12	0	0
46	1	1	0		11	0	0	0		44	0	0	0
	Original (a)					Quantized (b)					Reconstructed (c)		

Figure 57: Process of Quantization.

Let's stop here for a moment to understand and record our observations from this process.

1. The quantized numbers are smaller compared to the original numbers.
2. Several numbers have become zeros.
3. The *quantizer* value 4 controls the nature of the resulting number set. The higher this value, the more pronounced will be the above effects from observations 1 & 2.
4. Information was lost during this process and the original numbers were non-retrievable.

These observations are important as they form the principles by which significant compression is achieved in every encoder. In the following section we shall explore in detail how this is done.

7.3.2   QUANTIZATION MATRIX

As we've seen earlier, it is easy to perform quantization using a single quantization value. It's also possible to take it a step further and use a set of QP values to do quantization, typically one value for each of the frequency components. What this provides is the ability to leverage the HVS. It is more sensitive to changes in low frequency components when compared to the high frequency components. We can then optimize and customize the quantization process such that we employ higher quant values to discard higher frequency components and prioritize and preserve low frequency components using lower quantization values.

This set of quantization values is called a *quantization matrix* or *quantization table*. Furthermore, we can define different matrices for luma and chroma with higher quant values for chroma compared to luma in accord with the HVS.

Coming back to our earlier original number set. Let us assume this set has low frequency components at the start and higher frequency components toward the end as is typical in transformed blocks of video content.

Let us now define a quantizer set as shown in Figure 58 instead of a fixed value.

2	2	3	5
2	2	3	5
3	3	5	6
5	6	8	9

Figure 58: Quantization matrix.

Dividing the original number set that is replicated in Figure 59a by this quantizer set and discarding the remainders assuming this is an integer set, we get the following quantized 4x4 matrix in 59b. When we do the reverse operation of multiplication at the receiving end, we obtain the following reconstructed matrix of 58c. Clearly, the reconstructed values at the decoder end for the numbers at the top left of the 4x4 matrix, which correspond to the low frequency components, have improved compared to the results provided by a fixed quant value division.

While these numbers have become bigger with a wider range, this can be compensated for by more aggressive quantization values for the numbers down the series.

240	123	42	4		120	61	14	0		240	122	42	0
239	51	12	3		119	25	4	0		238	50	12	0
81	13	3	0		27	4	0	0		81	12	0	0
46	1	1	0		9	0	0	0		45	0	0	0

Original (a)   Quantized (b)   Reconstructed (c)

Figure 59: Quantization using a quantization matrix.

7.3.3   QUANTIZATION IN VIDEO

Now that we have understood the mechanics of quantization, let us explore how this is applied specifically to video compression. In compression, quantization is the next step that a block of residual samples undergoes after the transform process. We discussed earlier that the block of transform coefficients has the values concentrated around the low frequency components. We've also discussed how, during quantization, signals with higher range are mapped to become signals with a reduced range that need fewer bits for representation. When applied to the residual samples of video, quantization thus serves to reduce the precision of the remaining non-zero coefficients. Furthermore, it usually leaves us with a block in which most or all coefficients are zero. In video coding standards, two quantization indicators are important:

1. Quantization Parameter (QP): This is the application layer parameter that can be used to specify the level of quantization needed.
2. Quantization Step Size (Q_{step}): This is the actual value by which the transformed values are quantized by. Additional scaling may or may not be also applied along with the Q_{step} division process.

QP and Q_{step} are usually mathematically related and one can be derived from the other. For example, in the H.264 standard, QP values range from 0 to 51 and QP and Q_{step} are mathematically linked by the equation: $Q_{step} = 2^{((QP-4)/6)}$. In this scenario, every increase in QP value by 6 correspondingly doubles the value of Q_{step}. It then is obvious that by setting the QP to a high value, the Q_{step} is correspondingly increased and more coefficients become zero and vice versa. This means a higher QP results in lesser values to process and hence more compression at the expense of video quality.

Conversely, setting the QP to a low value leaves us with more non-zero coefficients, resulting in higher quality but a lower compression ratio. Dialing up or down the quantization parameter helps in spreading the allocation of bits across areas of the video, both within the frame and within a time interval across several frames. The challenge, thus, for encoder implementations is to arrive at suitable QP values. These provide the highest compression efficiency while maintaining the best visual quality possible. The QP thus becomes the most critical parameter in tuning the picture quality. We will explore in detail how this is done in the section on rate control.

The quantizer is applied to the transformed coefficients (T_{coeff}) as follows:

$$Q_{coeff} = \textbf{round } (T_{coeff}/Q_{step})$$

where Q_{step} is the quantizer step size.

-1	-1	-1	-1	-1	-1	-1	-1	-3	-3	-3	-3	-3	-3	-3	-3
0	0	-1	-1	-1	-1	-1	-1	-3	-3	-3	-3	-3	-3	-3	-3
1	0	-1	-1	-2	-2	-2	-2	-3	-3	-3	-3	-3	-3	-3	-3
1	1	0	-1	-2	-2	-3	-3	-3	-3	-3	-3	-3	-3	-3	-3
2	2	1	1	0	-1	-2	-2	-3	-3	-3	-3	-3	-3	-3	-3
2	2	1	1	0	0	-1	-2	-2	-2	-3	-3	-3	-3	-3	-3
0	0	1	1	0	0	-1	-1	0	-1	-2	-2	-3	-3	-3	-3
-3	-2	0	0	1	1	0	0	1	1	0	-1	-2	-2	-3	-3
-3	0	0	0	1	-1	-1	-1	-1	0	-1	0	0	0	0	1
-5	-2	-1	1	1	0	1	0	0	0	0	1	0	1	1	0
-7	-3	-1	0	0	0	1	1	-1	0	1	1	0	1	0	0
-10	-6	-5	-3	-3	-3	-2	-1	-2	-1	0	0	1	0	-1	-1
-13	-10	-8	-7	-8	-7	-7	-5	-5	-2	-1	-1	0	-1	-1	-1
-13	-11	-11	-11	-12	-12	-12	-9	-7	-5	-3	-2	-2	-2	-1	-2
-11	-10	-10	-12	-14	-15	-15	-12	-13	-9	-6	-6	-5	-4	-4	-3
-9	-9	-11	-14	-17	-19	-20	-18	-20	-16	-14	-11	-10	-9	-8	-6

Figure 60: A 16x16 block of residual values after prediction.

As an example, let us look at a 16x16 block of residual samples that's transformed using a 16x16 integer transform and quantized as shown in figures 60-62. In Figure 62, the smaller coefficients have become zero in the quantized block and the non-zero values are concentrated around the top-left coefficients that correspond to the low frequency components.

Furthermore, the non-zero coefficients in Figure 62 have a reduced range that now allows us to represent the quantized block values with fewer bits than the original signal. It should be noted that quantization is an *irreversible process*, meaning there is no way to exactly reconstruct the signal input from the quantized values. To illustrate this concept, let us now perform the reverse operations just as the decoder would do when it receives the quantized signal in Figure 62.

-399	-59	36	7	-32	-24	-12	-17	-6	2	-2	-13	-3	1	-3	-13
280	216	-35	-21	28	26	12	16	6	-1	3	10	0	-3	-1	6
-306	-61	98	39	15	13	4	-9	2	10	1	-7	6	10	-5	-8
174	-57	-96	-44	-20	3	-10	-8	-5	5	-3	-8	-5	-3	1	2
-67	31	36	0	8	-3	5	-1	4	-3	-1	-1	-1	1	11	0
24	-35	-54	2	-2	-9	-2	6	-1	-5	2	7	4	3	-5	3
-7	57	22	-16	-3	10	0	-8	0	5	2	-3	-1	-2	1	-4
25	-25	7	1	-2	-5	-1	1	-2	-1	-4	0	-4	-2	1	2
-30	5	8	0	-5	-1	1	-1	2	-6	2	-6	-1	6	0	-1
19	-6	-24	6	0	-4	-2	-3	-4	3	-3	0	8	1	1	0
-24	-3	6	1	3	-3	-1	2	1	0	0	12	-2	-2	0	1
8	-6	10	2	2	1	1	1	0	-3	3	-2	-3	1	-1	1
-9	-1	4	3	2	1	3	-1	-1	-1	-2	1	-1	0	0	-1
9	1	-17	-4	-5	0	3	0	-1	6	5	1	4	1	-1	-2
-6	0	0	2	1	3	-2	4	7	3	-3	3	-2	-3	-2	4
4	2	12	-2	2	-3	0	-5	3	3	0	-3	3	4	2	-2

Figure 61: The 16x16 block after undergoing a 16x16 transform.

-3	0	0	0	0	0	0	0	0	0	0	0	0	0	0	0
2	1	0	0	0	0	0	0	0	0	0	0	0	0	0	0
-2	0	0	0	0	0	0	0	0	0	0	0	0	0	0	0
1	0	0	0	0	0	0	0	0	0	0	0	0	0	0	0
0	0	0	0	0	0	0	0	0	0	0	0	0	0	0	0
0	0	0	0	0	0	0	0	0	0	0	0	0	0	0	0
0	0	0	0	0	0	0	0	0	0	0	0	0	0	0	0
0	0	0	0	0	0	0	0	0	0	0	0	0	0	0	0
0	0	0	0	0	0	0	0	0	0	0	0	0	0	0	0
0	0	0	0	0	0	0	0	0	0	0	0	0	0	0	0
0	0	0	0	0	0	0	0	0	0	0	0	0	0	0	0
0	0	0	0	0	0	0	0	0	0	0	0	0	0	0	0
0	0	0	0	0	0	0	0	0	0	0	0	0	0	0	0
0	0	0	0	0	0	0	0	0	0	0	0	0	0	0	0
0	0	0	0	0	0	0	0	0	0	0	0	0	0	0	0
0	0	0	0	0	0	0	0	0	0	0	0	0	0	0	0

Figure 62: The 16x16 block after undergoing quantization.

Figure 63 shows the 16x16 block after the process of inverse quantization. This is followed by Figure 64 that shows the final reconstructed residual values at the decoder after inverse transform. While these values clearly show patterns and are a fair approximation of the original 16x16 residual block in Figure 60, it's nowhere near an identical representation of the input source.

-387	0	0	0	0	0	0	0	0	0	0	0	0	0	0	0
322	161	0	0	0	0	0	0	0	0	0	0	0	0	0	0
-322	0	0	0	0	0	0	0	0	0	0	0	0	0	0	0
161	0	0	0	0	0	0	0	0	0	0	0	0	0	0	0
0	0	0	0	0	0	0	0	0	0	0	0	0	0	0	0
0	0	0	0	0	0	0	0	0	0	0	0	0	0	0	0
0	0	0	0	0	0	0	0	0	0	0	0	0	0	0	0
0	0	0	0	0	0	0	0	0	0	0	0	0	0	0	0
0	0	0	0	0	0	0	0	0	0	0	0	0	0	0	0
0	0	0	0	0	0	0	0	0	0	0	0	0	0	0	0
0	0	0	0	0	0	0	0	0	0	0	0	0	0	0	0
0	0	0	0	0	0	0	0	0	0	0	0	0	0	0	0
0	0	0	0	0	0	0	0	0	0	0	0	0	0	0	0
0	0	0	0	0	0	0	0	0	0	0	0	0	0	0	0
0	0	0	0	0	0	0	0	0	0	0	0	0	0	0	0
0	0	0	0	0	0	0	0	0	0	0	0	0	0	0	0

Figure 63: The 16x16 block after inverse quantization.

1	1	1	1	0	0	-1	-1	-2	-2	-2	-3	-3	-3	-4	-4
1	1	1	0	0	0	-1	-1	-2	-2	-3	-3	-3	-4	-4	-4
1	0	0	0	0	-1	-1	-1	-2	-2	-3	-3	-3	-4	-4	-4
0	0	0	0	-1	-1	-1	-2	-2	-2	-3	-3	-3	-3	-4	-4
0	0	0	0	-1	-1	-1	-1	-2	-2	-2	-3	-3	-3	-3	-3
0	0	0	0	0	-1	-1	-1	-1	-1	-2	-2	-2	-2	-2	-2
0	0	0	0	0	0	0	0	0	-1	-1	-1	-1	-1	-1	-1
1	1	1	1	0	0	0	0	0	0	0	0	0	0	0	0
0	0	0	0	0	1	1	1	1	1	1	1	1	1	1	1
0	0	0	0	0	0	0	0	0	0	1	1	1	1	1	1
-2	-2	-2	-2	-2	-2	-1	-1	-1	-1	0	0	0	0	0	0
-5	-5	-4	-4	-4	-4	-3	-3	-3	-3	-2	-2	-2	-2	-2	-1
-8	-7	-7	-7	-7	-6	-6	-5	-5	-5	-5	-4	-4	-4	-4	-4
-11	-10	-10	-10	-10	-9	-9	-9	-8	-8	-7	-7	-7	-6	-6	-6
-13	-13	-13	-12	-12	-12	-11	-11	-10	-10	-9	-9	-9	-8	-8	-8
-14	-14	-14	-14	-13	-13	-12	-12	-12	-11	-11	-10	-10	-10	-9	-9

Figure 64: The reconstructed 16x16 block after inverse 16x16 transform.

The above operations were performed at higher QP, upwards of 40. This introduces significant quantization of the transformed values, as we observe in Figure 62.

Let us now illustrate how this QP value affects the results above by doing the same operations at around QP 30 and also at around QP 20. This is shown in Figures 65 and 66.

As we see in Figure 66, with lower QP values of around 20, the reconstructed residual values are almost identical to the input source residual values in Figure 60, except for some rounding differences.

```
-3  -2  -2  -1  -1   0   0  -1  -2  -3  -3  -4  -4  -4  -3  -3
-2  -2  -2  -1  -1  -1  -2  -2  -2  -3  -3  -3  -3  -2  -2  -2
-1  -1  -1  -1  -2  -2  -3  -3  -3  -3  -3  -2  -2  -1  -1  -1
 1   1   0  -1  -2  -2  -3  -3  -3  -3  -3  -2  -2  -2  -2  -2
 3   2   1   0  -1  -2  -2  -3  -3  -3  -3  -3  -3  -4  -4  -4
 2   2   1   0   0  -1  -1  -1  -1  -2  -2  -3  -4  -4  -5  -6
 1   0   0   0   0   0   0   0   0   0  -1  -1  -2  -3  -4  -5
-1  -1  -1  -1  -1   0   0   1   1   1   0   0  -1  -1  -2  -2
-2  -2  -1  -1   0   0   1   1   1   1   1   1   1   1   1   1
-2  -2  -1   0   0   1   1   1   1   1   1   1   1   1   2   2
-3  -3  -2  -1   0   1   1   1   1   0   0   0   1   1   1   2
-6  -6  -5  -3  -2  -1  -1   0   0   0   0   0   0   0   0   1
-10 -9  -8  -7  -6  -5  -4  -3  -3  -2  -2  -2  -1  -1  -1  -1
-12 -12 -11 -11 -10 -9  -8  -7  -7  -6  -5  -4  -4  -3  -3  -3
-12 -12 -12 -13 -13 -13 -13 -12 -11 -10 -8  -7  -6  -6  -5  -5
-11 -11 -12 -14 -15 -15 -15 -15 -14 -12 -11 -10 -8  -8  -7  -7
```

Figure 65: The reconstructed 16x16 block after inverse 16x16 transform in QP 30 case.

```
-1  -1  -1  -1  -1  -1  -1  -1  -3  -3  -3  -3  -3  -3  -3  -3
 0   0  -1  -1  -1  -1  -1  -1  -3  -3  -3  -3  -3  -3  -3  -3
 1   0  -1  -1  -2  -2  -2  -2  -3  -3  -3  -3  -3  -3  -3  -3
 1   1   0  -1  -2  -2  -3  -3  -3  -2  -3  -3  -3  -3  -3  -3
 2   2   1   1   0  -1  -2  -2  -3  -3  -2  -3  -3  -3  -3  -3
 2   2   1   1   0   0  -1  -2  -2  -2  -3  -3  -3  -3  -3  -3
 0   0   1   1   0   0  -1  -1   0  -1  -2  -2  -3  -3  -3  -3
-3  -2   0   0   1   1   0   0   1   1   0  -1  -2  -2  -3  -3
-3   0   0   0   1  -1  -1  -1  -1   0  -1   0   0   0   0   1
-5  -2  -1   1   1   0   1   0   0   0   0   1   0   1   1   0
-7  -3  -1   0   0   0   1   1  -1   0   1   1   0   1   0   0
-10 -6  -5  -3  -3  -3  -2  -1  -2  -1   0   0   1   0  -1  -1
-13 -10 -8  -7  -8  -7  -7  -5  -5  -2  -1  -1   0  -1  -1  -1
-13 -11 -11 -11 -12 -12 -12 -9  -7  -5  -3  -2  -2  -2  -1  -2
-11 -10 -10 -12 -14 -15 -15 -12 -13 -9  -7  -6  -5  -4  -4  -3
-9  -9  -11 -14 -17 -19 -20 -18 -19 -16 -14 -11 -10 -9  -8  -6
```

Figure 66: The reconstructed 16x16 block after inverse 16x16 transform in QP 20 case.

These examples establish the following observations that are of paramount importance to a video engineer.

1. The quantization process is irreversible and the signal transmitted and received after quantization processes is an approximate, lossy version of the original source.
2. Higher quantization results in loss of signal fidelity and therefore higher compression.

Control of quantization values is key to striking a balance between preserving signal fidelity and achieving a high compression ratio.

7.3.4   HOW CAN QP VALUES BE ASSIGNED?

Quantization is as much of an art as a science, as it involves analyzing the visual effects of discarding information. Significant research has been done on applying subjective quantization matrices to get the best visual experience at the highest compression efficiency. Novel ideas are explored wherein a unique matrix or QP value can be signaled at every frame level, differently for luma and chroma, or even signal a different matrix, depending on frame type in the GOP.

Figure 67: Effects of quantization process.
source video: https://media.xiph.org/video/derf/

With basic signaling at the frame level in place, the resulting QP values can also be adjusted within the picture frame, depending on complexity, with flatter areas getting lower QP adjustments, thereby preserving details at the expense of complex areas. This is explained in the adaptive quantization section in chapter 10. Figure 67 shows an example of a section of a video that's encoded with different quantization values. The lower QP value, say, QP 30 as shown in the top left corner of the image preserves the fidelity of the samples better and we can easily see the details of the tree leaves.

As the QP increases, however, the encoded video becomes dramatically different from the source, resulting in quantization blocky artifacts, as observed in the bottom image that is encoded with QP 50. Modern encoders have built-in advanced spatial and temporal adaptive algorithms that analyze the scene content to calculate and optimally allocate the QP at a block-level based on scene complexity. This helps to avoid unpleasant blocky effects in areas of detail and thereby provides significant visual quality benefits.

7.4   REORDERING

4	2	2	1	-1	1	0	0
2	2	1	-1	0	0	0	0
-2	1	1	1	0	0	0	0
1	1	0	0	0	0	0	0
1	0	0	0	0	0	0	0
0	0	0	0	0	0	0	0
0	0	0	0	0	0	0	0
0	0	0	0	0	0	0	0

Figure 68: 8x8 block of quantized coefficients.

As we have seen, the quantized transform coefficients have several zero coefficients and the non-zero coefficients are concentrated around the top left corner. Instead of transmitting all the values, which invariably include redundant zeros, it becomes beneficial to transmit the very few coefficient values and signal the remaining values as zeros. The decoder, when it receives the bitstream, would be able to derive the non-zero coefficients and then use the zero signaling to add zeros to the remaining values.

In order to do this efficiently, it's necessary that the non-zero coefficients be all packed together and the zero values be packed together separately. The best possible packing is then the one in which a maximum number of zeros, also called *trailing zeros*, are packed toward the end.

This can then be signaled just once. As an example, let us consider the 8x8 transformed and quantized coefficients as shown in Figure 68. Parsing the above block horizontally in a simple raster order produces the following stream of values:

[4 2 2 1 -1 1 0 0 2 2 1 -1 0 0 0 0 -2 1 1 1 0 0 0 0 1 1 0 0 0 0 0 1 (and thirty-one 0s)]

By scanning the block horizontally or vertically, we see that a lot of zeros are still present in between non-zero coefficients, resulting in an inefficient representation. This because the frequency organization of the 8x8 block is not horizontal or vertical but instead a different pattern.

By using the pattern of frequency organization, we are able to organize the coefficients in the order of increasing frequencies. When this order is used, the non-zero coefficients that pertain to the low frequency components are concentrated at the start, followed by high frequency components. These are usually zeros after quantization process.

The pattern of frequency organization is called *zig-zag scan* and is illustrated in Figure 69, below. When the previous residual block in Figure 68 is reordered using the zigzag scan pattern, starting at the top-left and traversing the block in a zigzag manner to the bottom right, it produces the following list of coefficients.

[4, 2, 2, -2, 2, 2, 1, 1, 1, 1, 1, 1, 1, -1, -1, 1, 0 ,1, (and forty-six 0s)]

We notice that the number of trailing zeros is significantly higher than the raster scan pattern. The result is more efficient encoding. However, we also notice that there are still zeros between the coefficients that could potentially be further optimized.

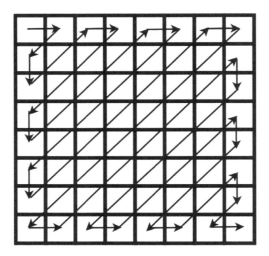

Figure 69: Zig-zag scanning order of coefficients of 8x8 block.

Scan tables like the classic zig-zag scan shown in Figure 69, thus, provide a much more efficient parsing of coefficients. All non-zero coefficients are grouped together first, followed by zero coefficients. Other scan tables are also possible and in fact VP9 provides a few different pattern options that organize coefficients more or less by their distance from the top left corner. The default VP9 scan tables, showing the order of the first few coefficients, is shown in Figure 70, below.

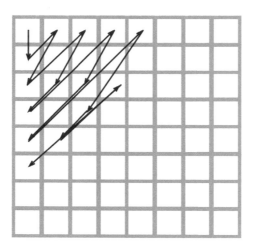

Figure 70: Default scanning order of coefficients of 8x8 block in VP9.

As we see below, the earlier 8x8 block of quantized coefficients can be also efficiently represented when the VP9 scanning pattern shown in Figure 70 is used for reordering.

[4, 2, 2, -2, 2, 2, 1, 1, 1, 1, 1, 1, 1, -1, -1, 1, 0 ,1, (and forty-six 0s)]

It should be noted that the scan pattern is very much tied to the underlying transform. It removes redundancies among residual samples and provides energy compaction. As VP9 provides flexible combinations of DCT and ADST for horizontal and vertical transforms, it also provides flexible scanning options which can be used in conjunction with the transform combinations.

7.5   RUN LEVEL PAIR ENCODING

Run level pair encoding is a technique to efficiently signal the large number of zero values in the quantized block. The basic concept is that, instead of encoding all the zeros that exist among the coefficients individually, which consumes bits, the number of zeros is signaled in the bitstream using a single value.

Specifically, the number of leading zeros before any non-zero coefficient is signaled in the bitstream. This can save some bits, especially if there are several zeros between successive non-zero coefficients. Let us now explore how the earlier array can be broken down using this concept.

Sequence of numbers before run level encoding:

[4, 2, 2, -2, 2, 2, 1, 1, 1, 1, 1, 1, 1, -1, -1, 1, 0 ,1, (and forty-six 0s)]

Run level encoding strategy:

[(zero 0s followed by 4), (zero 0s followed by 2), (zero 0s followed by 2), (zero 0s followed by -2), (zero 0s followed by 2), (zero 0s followed by 2), (zero 0s followed by 1), (zero 0s followed by 1), (zero 0s followed by 1), (zero 0s followed by 1), (zero 0s followed by 1), (zero 0s followed by 1), (zero 0s followed by 1), (zero 0s followed by -1), (zero 0s followed by -1), (zero 0s followed by 1), (one 0 followed by 1), forty-six 0s]

Sequence of numbers after run level encoding:

[(0,4)(0,2)(0,2)(0,-2)(0,2)(0,2)(0,1)(0,1)(0,1)(0,1)(0,1)(0,1)(0,1)(0,-1)(0, 1)(0,1)(1,1)(End of block)]

The first number in every (*run*, *level*) pair, namely, *run*, indicates the number of immediately preceding zeros and the second number, namely, *level*, indicates the non-zero coefficient value. The *end of block* is a special signal that can be communicated in the bitstream to indicate *no more bits* and all else are zeros. Another way of organizing the same information could be assigning a value to every symbol that indicates the end of block. The previous sequence of numbers would then look as follows:

[(0,4,0)(0,2,0)(0,2,0)(0,- 2,0)(0,2,0)(0,2,0)(0,1,0)(0,1,0)(0,1,0)(0,1,0)(0,1,0)(0,1,0)(0,1,0)(0,- 1,0)(0,-1,0)(0,1,0)(1,1,1)]

Notice that the last value is set to 1, indicating the end of block. When the decoder receives this stream, it looks for the last 1-bit and sets the remaining values of the block to zeros.

In the next chapter we shall see how the compactly represented, quantized signal gets encoded as bits and bytes in the final encoded bitstream.

7.6   SUMMARY

- Transforms take in a block of residual pixel values (after prediction) and convert them to the frequency domain. This amounts to the same values being represented differently.

- Pixel values vary in intensity and the time it takes to change from one intensity to another and back again is represented by frequency. The faster the change of intensity from, say, light to dark and back, the higher the frequency needed to represent that part of the picture.
- An image pixel block can be broken down into a linear weighted sum of its constituent frequency components with higher frequency components representing more details.
- Transforms provide energy compaction. This is the fundamental criterion in their selection.
- The DCT is widely used in video coding standards as it provides a high degree of energy compaction.

- Quantization is the process of reducing the range of the set of transformed residual values. It can be achieved by division at the encoding side and a corresponding multiplication at the decoding side.
- Quantization is an irreversible process.
- Higher quantization results in loss of signal fidelity. Higher compression and the control of quantization values is key to striking a balance between preserving signal fidelity and achieving a high compression ratio.

8 ENTROPY CODING

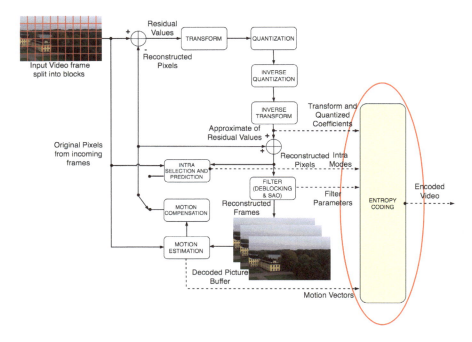

In the previous chapters we explored how inter and intra pixel redundancies are removed to minimize the information that needs to be encoded. We also saw mechanisms to efficiently represent the resulting residuals using transforms, quantization, scanning and run-level coding.

The following pieces of information have to be sent as part of the bitstream at a block level: *motion vectors, residuals, prediction modes, and filter settings information*. In this chapter, we will study in detail how the run-level values are encoded using the fewest bits by minimizing statistical coding redundancies. Recall that, in the previous section, we had the following run level pairs for the example encoding block:

[(0,4,0)(0,2,0)(0,2,0)(0,-2,0)(0,2,0)(0,2,0)(0,1,0)(0,1,0)(0,1,0)(0,1,0)(0,1,0)(0,1,0)(0,1,0)(0,-1,0)(0,-1,0)(0,1,0)(1,1,1)]

The easiest way to encode this to a binary bitstream would be to understand the range of all possible values of such symbols, then determine the number of bits needed to encode them with a fixed number of bits per symbol. Assuming there are 100 such symbols each for last = 0 and last = 1, then we can create a unique value using 8 bits for every symbol. To encode the above set we would then need 17 x 8 = 136 bits or 17 bytes.

This implicitly assumes that all the symbols have the same likelihood of occurrence, in which case assigning the same number of bits to each symbol makes sense. However, in reality, data including video and image content rarely have symbols that are equally likely. Instead, they tend to have some symbols that occur more frequently than others.

8.1   INFORMATION THEORY CONCEPTS

Information theory helps us understand how best to send content with minimum bits for symbols with unequal likelihoods. This was presented first by Claude Shannon in his landmark paper, mentioned in Chapter 3. Shannon showed clearly the limits on how much data can be compressed in a lossless manner. In this section, we try to present intuitively how it works and also explain mathematically how the concepts are formulated.

Let's start by illustrating with a simple example where English sentences are being communicated and need to be compressed. In this case, the symbols could be any of the letters of the English alphabet and the number of symbols thus is 26. Let's say the word that needs to be sent is "seize." When we receive the letter 's', it's a characteristic of the language that the next letter is more likely (higher probability) to be a vowel than say a letter 'b' (low probability).

As we already know we're more likely to receive a vowel next, receiving a vowel thus has *less information* to us than receiving a letter like 'b'. Intuitively, we can thereby see the relationship between likelihood of occurrence (probability) and information. Information theory tells us the exact same principle expressed mathematically. Let's say we have the symbol set $\{x_1, x_2, x_3 ... x_n\}$. Let $P(x_m)$ represent the probability of symbol x_m occurring in the absence of other information. Then,

$$P(x_1) + P(x_2) + P(x_3) ... + P(x_n) = 1$$

$$INF(x_m) = -\log_2 P(x_m) \text{ bits}$$

In other words,

$$\text{INF}(x_m) = \log_2 [1 / P(x_m)] \text{ bits}$$

where $\text{INF}(x_m)$ is the information of x_m in bits.

Let us illustrate the above using a simple example. Let's again come back to our earlier example of sending English text and the word "seize." Assuming for a moment the symbols are limited to the set {'s', 'e', 'i', 'z'}, their probabilities of occurrence are as follows:

$$P(s) = \tfrac{1}{5}, P(e) = \tfrac{2}{5}, P(i) = \tfrac{1}{5}, P(z) = \tfrac{1}{5}$$

Therefore,

$$\text{INF}(s) = \text{INF}(i) = \text{INF}(z) = -\log_2(1/5) = 2.32$$

$$\text{INF}(e) = -\log_2(2/5) = 1.32$$

In this simple example, we see that receiving the letter 'e', because it has higher probability of occurrence, has less information than receiving the letters 's' or 'i' or 'z'.

The information of any symbol is inversely proportional to the likelihood of occurrence of the symbol. The higher the likelihood, the less information carried by the symbol and hence the fewer should be the number of bits needed for that symbol and vice versa. This is the philosophy employed in entropy coding schemes like *variable length coding* scheme or the popular *binary arithmetic* coding used extensively since the advent of H.264. We can also see this in our example sequence where the symbol (0, -1, 0) occurs 8 times and (0,2,0) occurs three times. In such a scenario wherein, the symbols are not equally likely, we now know it's best to not assign equal bits across all symbols and instead strategically assign the bits such that *the most-occurring symbols get the fewest bits and the least occurring symbols get the most bits.* This will still ensure a unique symbol-to-bits mapping while optimizing the number of bits encoded in the bitstream.

8.1.1   THE CONCEPT OF ENTROPY

The term *entropy* is widely used in information theory and can be intuitively thought of as a measure of randomness associated with a specific content. As we have seen earlier, the more random the content or its associated symbols, the more information it has and thereby more bits are needed for its transmission and vice versa.

> Entropy thus is simply the average amount of information from the content and measures the average number of 'bits' needed to express the information.

Mathematically, entropy is expressed by the formula:

$$H(X) = -\sum_m P(x_m) \log_2 P(x_m)$$

H () represents entropy, which is the average number of bits.

X represents the set of all values in the content.

x_m represents a symbol of the set X.

Coming back to our previous example, where we calculated the information for every symbol using the formula: INF (x_m) = -$\log_2 P(x_m)$ wherein:

$$INF(s) = INF(i) = INF(z) = -\log 2(1/5) = 2.32$$

$$INF(e) = -\log 2\ (2/5) = 1.32$$

The *average information* or *entropy* in this simple example would be a weighted average of the information of all the associated symbols. Mathematically,

$$\text{Entropy } H\ (X) = -\sum_m P(x_m) * \log_2 P(x_m)$$

$$H\ (X) = \tfrac{1}{5} * 2.32 + \tfrac{1}{5} * 2.32 + \tfrac{1}{5} * 2.32 + \tfrac{2}{5} * 1.32 = 1.92$$

8.1.2   HOW ARE THE LIKELIHOODS OR PROBABILITIES HANDLED?

We now know that symbol likelihood-based entropy coding helps to optimize the number of bits sent in the bitstream. It's also clear that entropy or the information in the data makes sense only when discussed in relation to their associated probability distribution. However, how are these probability distributions determined? In the above example, we computed the probabilities for one word with the symbol set, 's', 'e', 'i', and 'z'. If we had an extended library of all words with this symbol set, we could learn from it and improve the probabilities. The same principle can be extended to the symbols that are encoded in image and video, where in the probabilities are usually calculated by running several encoding simulations across a wide library of video content and counting the number of occurrences of every symbol. Such simulations are done at the time of standardization and the base probability tables for every symbol that is encoded are usually included in the normative part of the standard.

As content and complexity within the video sequence is constantly changing with every frame, so are the corresponding symbol likelihoods. How could the symbol likelihoods then be adaptively determined under such dynamic conditions? This is the challenge that's addressed by *context adaptive entropy coding algorithms*. These keep running counts of the recurrences of every symbol during the process of encoding. The algorithms use the counts to update the probabilities (called context), either every CTU or macroblock as in H.264 and H.265, or at the end of every frame as in VP9.

As the statistics of symbols like motion vectors and residuals vary spatially and temporally and also across bit rates, and so on, adapting the statistics based on already coded symbols is recommended. The philosophy here is that, by constantly updating the probabilities as scenes and pictures or settings change, better allocation of bits for the symbols based on updated occurrences leads to better coding efficiency. H.264 defined two context adaptive entropy encoding schemes, namely, *context adaptive variable length coding* (CAVLC) and *context adaptive binary arithmetic coding* (CABAC). However, newer standards like H.265 and VP9 only support CABAC and have slightly different methods in which the contexts are maintained and updated. In keeping with the newer trends and in an effort to keep the contents concise, this book will focus solely on CABAC.

8.2   CONTEXT ADAPTIVE BINARY ARITHMETIC CODING

As we've seen in the earlier section, there are two elements to modern entropy coding schemes, namely:

1. Coding algorithm (like arithmetic coding)
2. Context adaptivity

CABAC employs the above principles and has been found to achieve good compression performance through:

(a) selecting probability models for each syntax element according to the element's context,
(b) adapting probability estimates based on local statistics, and
(c) using arithmetic coding. [1][2]

Coding a data symbol using CABAC involves the following three stages that will be explained in detail in this section:

1. Binarization
2. Context modeling
3. Arithmetic coding

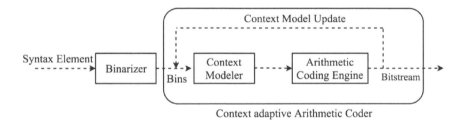

Figure 71: Block diagram of context adaptive binary arithmetic coder.

Figure 71 illustrates the stages of this process. At the first step, if the input symbol is not binary-valued, it is mapped to a corresponding binary value in a process called *binarization*. The individual bits in the resulting binary value are called bins. Thus, instead of encoding the symbols themselves, we will focus on encoding their mapped binary equivalents.

In designing the binary value mapping for every symbol, care is taken to ensure no binary value pattern occurs within another binary value so as to ensure that every binary value received by the decoder can be uniquely decoded and mapped to an encoded symbol. The next step is to select a suitable model based on the past symbol distribution through a process called context modeling. The last step is the adaptive arithmetic encoding stage that adapts itself using the probability estimates that are provided by the earlier stages.

As the probability distribution of the input symbol is highly correlated to the probability distribution of the bins of its binary equivalent, the probability estimates of the neighboring symbols can be used to estimate fairly accurately the probabilities of the bins that will be encoded. After every bin is encoded, the probability estimates are immediately updated and will be used for encoding subsequent bins.

Probability distribution of Symbol ≈ Probability distribution of bins

8.2.1   BINARIZATION

Most of the encoded symbols, for example, residuals, prediction modes, and motion vectors, are non-binary valued. *Binarization* is the process of converting them to binary values before arithmetic coding. It is thus a pre-processing stage. It is carried out so that subsequently a simpler and uniform binary arithmetic coding scheme can be used, as opposed to an *m-symbol arithmetic coding* that is usually computationally more complex. It should be noted, however, that this binary code is further encoded by the arithmetic coder prior to transmission. The result of the binarization process is a binarized symbol string that consists of several bits. The subsequent stages of context modeling, arithmetic encoding and context updates are repeated for each bit of the binarized symbol string. The bits in the binarized string are also called bins. In H.265, the binarization schemes can be different for different symbols and can be of varying complexities.

In this book, we shall illustrate a few binarization techniques that have been employed in H.265. These include Fixed Length binarization technique and a concatenated binarization technique that combines Truncated Unary and Exp-Golomb binarization. The same concepts can be extended to other schemes.

8.2.1.1   FIXED LENGTH (FL) BINARIZATION

This is a simple binarization scheme wherein the binarization string for the symbol corresponds to the actual binary representation of the symbol value.

If x denotes a syntax element of a finite set such that $0 \leq x < S$, we first determine the minimum number of bits needed to represent the range of values of the symbol set. This is given by:

$$l_{FL} = \log_2(S+1)$$

The FL binarization string of x is then simply given by the binary representation of x with l_{FL} bits. Table 10 illustrates this with a simple example for a symbol that takes values from 0 to 7.

Here, $S = 7$ and hence $l_{FL} = \log_2 8 = \mathbf{3}$ bits.

As this is a fixed length representation, it's natural to apply it to symbol sets that have a uniform distribution.

Table 10: Binary codes for fixed length (FL) binarization.

x	$B_{FL}(x)$		
0	0	0	0
1	0	0	1
2	0	1	0
3	0	1	1
4	1	0	0
5	1	0	1
6	1	1	0
7	1	1	1

8.2.1.2 CONCATENATED BINARIZATION

The next binarization technique we shall describe is a concatenation of two basic schemes, namely, *truncated unary* (TU) technique and *exp-Golomb* technique. The concatenated technique, called *unary/kth order exp-Golomb*

(UEGk) binarizations, is applied to MVDs and transform coefficient levels in H.265.

The unary code is the simplest prefix-free code to implement and easy for context adaptation. However, as larger symbol values don't really benefit from context adaptation, the combination of a truncated unary technique as a prefix and a static exp-Golomb code as a suffix is used.

In the following section, we describe each of these techniques and also how they are used in concatenation.

8.2.1.3 TRUNCATED UNARY BINARIZATION TECHNIQUE

The TU coding scheme is similar to the *unary coding* scheme that is used to represent non-negative numbers. In this scheme, any symbol x belongs to a symbol set with S values such that $0 \leq x \leq S$ is represented by **x** '1' bits and an extra '0' termination bit.

Table 11: Binary codes for TU binarization.

x	$B_{TU}(x)$							
0	0							
1	1	0						
2	1	1	0					
3	1	1	1	0				
4	1	1	1	1	0			
5	1	1	1	1	1	0		
6	1	1	1	1	1	1	0	
7	1	1	1	1	1	1	1	0
8	1	1	1	1	1	1	1	1

The length of this transformed bin string is thus x + 1. In TU, for x = S, the termination bit is not used and only x '1' bits are used, resulting in S bits for the maximum value.

Table 11 illustrates this with a simple example for a symbol which takes values from 0 to 8. Here, S = 8 and hence l_{TU} = **8** bits.

8.2.1.4 EXP-GOLOMB BINARIZATION TECHNIQUE

A *k-th order exp-Golomb* code (EG_k) is derived by using a combination of unary code that is used as a prefix and padded with a suffix having the following variable length (l_s):

$$l_s = k + l_p - 1$$

where l_p is the length of the unary prefix code.

Table 12: Binary codes for 0th and 1st order exp-Golomb binarization code.

x	l_{EG0}	Unary				Suffix				x	l_{EG1}	Unary				Suffix			
0	1				0					0	2					0	0		
1	3			1	0	0				1	2					0	1		
2	3			1	0	1				2	4			1	0	0	0		
3	5		1	1	0	0	0			3	4			1	0	0	1		
4	5		1	1	0	0	1			4	4			1	0	1	0		
5	5		1	1	0	1	0			5	4			1	0	1	1		
6	5		1	1	0	1	1			6	6	1	1	0	0	0	0		
7	7	1	1	1	0	0	0	0		7	6	1	1	0	0	0	1		
8	7	1	1	1	0	0	0	1		8	6	1	1	0	0	1	0		
9	7	1	1	1	0	0	1	0		9	6	1	1	0	0	1	1		

In general, k-th order EG_k code has the same prefix across different k values but varies in suffix by a factor of k. Every k-th order EG_k code scheme starts with k suffix bits and progresses from there. Examples for EG_k codes for k=0 and k=1 are given in Table 12. EG_0 code schemes start with 0 bits for suffix for their first value x=0 and then add 1 bit for suffix for x=1,2 and so on. In contrast, EG_1 schemes start with 1-bit suffix code for x=0,1 and so on. Now that we know how TU and EG_k codes work, let's conclude this section with an example of a UEG_k binarization code scheme that involves a simple concatenation of TU and EG_k codes.

Table 13: Binary codes for UEG_0 binarization.

x	$B_{TU}(x)$ (PREFIX)	EG_k (SUFFIX)
0	0	
1	1 0	
2	1 1 0	
3	1 1 1 0	
4	1 1 1 1 0	
:	: : : : :	
13	1 1 1 1 1 1 1 1 1 1 1 1 1 0	
14	1 1 1 1 1 1 1 1 1 1 1 1 1 1 0	
15	1 1 1 1 1 1 1 1 1 1 1 1 1 1 1 1	0
16	1 1 1 1 1 1 1 1 1 1 1 1 1 1 1 1	1 0 0
17	1 1 1 1 1 1 1 1 1 1 1 1 1 1 1 1	1 0 1
18	1 1 1 1 1 1 1 1 1 1 1 1 1 1 1 1	1 1 0 0 0
19	1 1 1 1 1 1 1 1 1 1 1 1 1 1 1 1	1 1 0 0 1
20	1 1 1 1 1 1 1 1 1 1 1 1 1 1 1 1	1 1 0 1 0
21	1 1 1 1 1 1 1 1 1 1 1 1 1 1 1 1	1 1 0 1 1

As illustrated in Table 13, the scheme uses a TU prefix with a truncation cut-off value S = 14 and EG_k suffix of order k=0.

Different schemes of this type are deployed in H.265 for different symbols like MVDs and transform coefficients. These schemes vary primarily in the cut-off points for the TU scheme and also the order k of the EG_k suffix. These values are chosen after a careful consideration of the typical magnitudes of these symbols and their probability distributions.

VP9 video standard employs a very similar design framework with slight differences in implementation choices and terminology. Every non-binary symbol is binarized to construct a binary tree and each internal node of the binary tree corresponds to a bin. The tree is traversed, and the binary arithmetic coder is run at each node to encode a particular symbol. Every node (bin) has an associated probability with 8-bit precision and this set of probabilities of the nodes is the maintained context for the symbol. Now that we've understood how the non-binary symbols are binarized to produce a binary bin-stream, let us delve into the details of the next step: context modeling.

8.2.2   CONTEXT MODELING

Each bin has an associated probability of being a '1' or '0'. This is determined by its context model. One model is chosen for every bin from many available models based on the statistics of previous coded symbols. Figure 72 is extracted from the earlier Figure 71. It shows how context modeling is closely tied to the arithmetic encoding stage. The model tells the binary encoder what the probability of a bin is in an input bin string. If the model provides an accurate probability estimation of the bins, they will be encoded optimally.

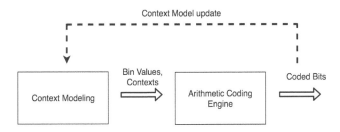

Figure 72: Context modeling and arithmetic Coding.

If, on the other hand, inaccuracies exist in the model, this would result in a loss in coding efficiency. Thus, it's clear that the efficiency of arithmetic encoding is closely tied to the probabilities of the symbols or bins provided from the modeling stage and the arithmetic coder design provides a flexible context adaptation mechanism to update the probability distribution of the symbols or bins dynamically. This is further illustrated with an example in the section on arithmetic coding.

After encoding, the arithmetic coder analyzes the sequence of resulting coded bits and correspondingly updates the probability distribution (context model). These updated probabilities will be used by the model to provide input to encode the next input bin string. The updates to the model are critical in arithmetic coding but they come at the expense of higher complexity to both the encoder and the decoder. Different standards have chosen different strategies to strike a balance between the complexity and coding gains.

H.264 and H.265 use continual, per-symbol probability updates. This means that after each bit is encoded, the corresponding probability is updated. This is useful in intra coding particularly. However, these probabilities are not temporally carried forward but are reset every frame. This means that these codecs don't take advantage of coding redundancies between frames. VP9, on the other hand, does things differently by keeping the probabilities constant within the frame. This means that there is no per-symbol update after encoding every superblock. Instead, probabilities are updated after encoding every frame. This is done to keep the decoder implementation simpler. This mechanism of updating the contexts based on the symbol statistics of the previous frames without any explicit signaling in the bitstream is called *backward adaptation*. VP9 also provides a mechanism to be able to explicitly signal the probabilities in the header of each frame before it's encoded/decoded. This mechanism is called *forward updates*. By being able to adapt probabilities between frames, VP9 derives compression efficiency, especially during longer GOPs where several inter frames are used between key frames.

8.2.3   ARITHMETIC CODING

According to information theory, symbols with very high probability of occurrence tend to have average information, or entropy, of less than 1 bit. A primary advantage of arithmetic coding over previous schemes like

Huffman coding (which cannot encode less than one bit-per-symbol) is that arithmetic coding allows for encoding symbols at a fractional number of bits. This leads to improved compression efficiency, especially for video content, as this tends to have symbols with high correlation and hence high probabilities. While a fractional number of bits per symbol can seem a vague idea at first, it's again a matter of transform or change of perspective. Let's explore the idea and illustrate in this section how a fractional number of bits per symbol are even possible.

Note that the details of arithmetic coding are applicable for both binary and non-binary symbols. However, keeping in mind the importance of simplicity in implementation, video standards like H.265 and VP9 employ binary arithmetic coding and this will be the focus of this book.

8.2.3.1 ARITHMETIC CODING FUNDAMENTALS

In this section, I will explain first the core arithmetic coding idea and how it's able to achieve fractional bits per symbol. We will work through an example. The process of arithmetic coding involves what we can call a *transform operation*. The symbols to be encoded are transformed into coded number intervals and successive bins are transformed and coded by recursive interval sub-division. The outcome of this process is that a sequence of bins can be represented by a single fractional number in a defined fractional interval whose binary representation is then encoded in the bitstream.

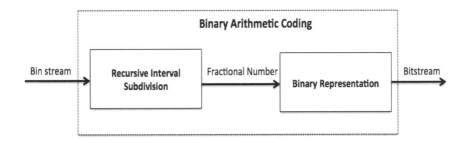

Figure 73: Process of binary arithmetic coding.

This means that no longer is the symbol bin stream directly encoded but, rather, its mapped fractional representation is instead encoded. It is this mapping process that facilitates encoding with better compression efficiency, such that the final bitstream has a fractional number of bits compared to the input bin stream. This process relies critically on the input probability context model and we will show how this is so shortly. The stages involved in the binary arithmetic coding process are illustrated in Figure 73.

Let us assume we have the following stream of 7 bits to be coded

[0 1 0 0 0 0 1] and $P_0 = 0.7$ and $P_1 = 0.3$

To start the process, the available interval range is assumed to be [0, 1] and the goal is to identify a final interval specific to the sequence of symbols and to pick any number from that interval. To do this, the binary symbols are taken one by one and assigned sub-intervals based on their probabilities, as illustrated in Figure 74 for our example bin stream.

The first bit is '0'. It has $P_0 = 0.7$ and is assigned the interval [0, 0.7]. This interval is chosen as the initial interval to encode the next bit. In this case, the next bit is '1' (with $P_1 = 0.3$). Based on this the interval [0,0.7] is further broken down into [0.49, 0.7]. This then becomes the initial interval for the next bit and so on.

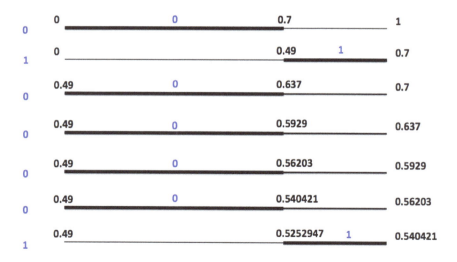

Figure 74: Illustration of coding a sample sequence using arithmetic coding.

The process can thus be summarized in the following 3 steps:

1. Initialize the interval to [0,1].
2. Calculate the sub-interval based on the incoming bit and its probability value.
3. Use the subinterval as the initial interval for the next bit and repeat step 2.

In our example, the final fractional interval is [0.5252947, 0.540421]. If we pick a number in this interval, say, 0.54, its binary equivalent is 0.10001. This can be then sent in the bitstream in the form of 5 bits: [10001]. It's clear from this example how a series of 7 input binary symbols can be compactly represented using just 5 bits, thereby achieving a fractional number of bits per symbol—0.714 bits per symbol in this case.

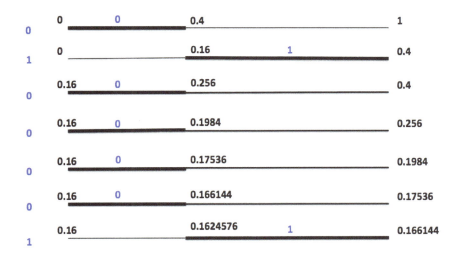

Figure 75: Coding the sample sequence using different context probabilities.

Now we'll explore how context probabilities affect this coding scheme using the same sequence of input bits but using different probabilities, say, $P_0 = 0.4$ and $P_1 = 0.6$. Figure 75 shows how this sequence will be coded. In this scenario, the final fractional interval is found to be [0.1624576, 0.166144]. If we pick a number in this interval, say, 0.165, its binary equivalent is 0.0010101. This can be effectively represented in the bitstream using a minimum of 7 bits. This is more than the 5 bits needed with probabilities $P_0 = 0.7$ and $P_1 = 0.3$. This clearly demonstrates the critical importance of accurate probability context models to provide coding gains using an arithmetic coding scheme.

However, arithmetic coding provides a flexible way to offset inaccurate probabilities by building in dynamic adaptation of probability estimates, based on the encoded symbols.

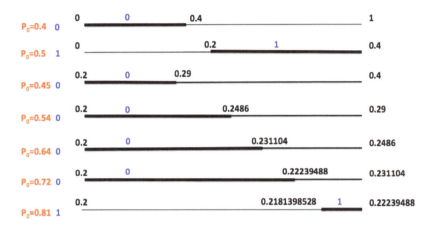

Figure 76: Illustration of dynamic probability adaptation in arithmetic coding.

In the above example with the initial $P_0 = 0.4$ and $P_1 = 0.6$, we can build in a simple mechanism to adapt the symbol probabilities at every stage, based on the bits encoded up to that stage. Then we can update the intervals accordingly, overcoming the limitation of the inaccurate initial probability. We can then represent the sequence of the same symbols using a number, say, 0.22, within the final interval [0.2181398528, 0.22239488] represented by [00111] using only 5 bits. This is illustrated in Figure 76, above. Figure 76 also highlights how P_0 is updated at every step to better reflect the symbol probabilities.

8.2.3.2 ARITHMETIC DECODING

Having explored how the arithmetic encoding works, let us now explore how its reverse operation, namely, arithmetic decoding works. This is illustrated in Figure 77 and is the reverse of the earlier binary encoding steps.

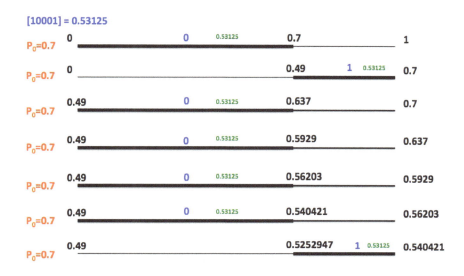

Figure 77: Illustration of decoding an arithmetic coded bitstream.

The interval range is initialized to [0,1] with known probabilities P0 = 0.7 and P1 = 0.3. These values are usually implicitly and dynamically computed by the decoder. When the decoder receives the binary sequence [10001], it interprets it as the fraction 0.10001 corresponding to decimal 0.53. The interval [0,1] can be successively sub-sectioned based on the context probabilities as follows.

Since 0.53125 lies between 0 and 0.7 the first symbol is determined as 0. Then the interval is set to [0,0.7] and the sub-intervals are [0,0.49] and [0.49,0.7] based on the P_0 value. Since 0.53125 now lies between 0.49 and 0.7 that corresponds to P_1 = 0.3 probability interval, the symbol is determined to be 1. This step is repeated in succession to arrive at the sequence of symbol bits, namely, [0 1 0 0 0 0 1] as shown in Figure 77.

The arithmetic coding process described in the previous section involves multiplication. This is avoided in CABAC implementations by approximating the interval values using lookup tables that are specified in the standard. This approach can potentially impact compression efficiency but is needed to keep the implementation simple. Also, during the internal encoding/decoding process, when the interval range drops below thresholds that are specified in the standard, a reset process is initiated. In this process, the bits from previous intervals are written to the bitstream and the process continues further.

8.3   SUMMARY

- Information theory helps us understand how best to send content with minimum bits for symbols having unequal likelihoods. Entropy coding is based on information theory.
- The term entropy is widely used in information theory and can be intuitively thought of as a measure of randomness associated with a specific content. In other words, it is simply the average amount of information from the content and measures the average number of 'bits' needed to express the information.
- Context adaptive entropy coding methods keep a running symbol count during encoding and use it to update probabilities (contexts) either at every block level or at the end of coding the frame. In doing so, the entropy coding context probability 'adapts' or changes dynamically as the content is encoded.
- CABAC is a context adaptive entropy coding method which includes the following functions in sequence: binarization, context modeling and arithmetic coding.
- Arithmetic coding involves a transformative operation where the symbols are transformed to coded number intervals and successive symbols are coded by recursive interval sub-division.

8.4   NOTES

1. Marpe D, Schwarz H, Wiegand T. Context-based adaptive binary arithmetic coding in the H.264/AVC video compression. Oral presentation by C-H Huang at: IEEE CSVT meeting; July 2003. https://slideplayer.com/slide/5674258/ Accessed September 22, 2018.

9 FILTERING

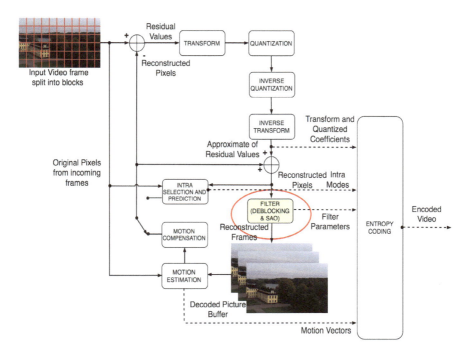

A deblocking or in-loop filter is applied to the decoded (reconstructed) pixels to remove the artifacts around block boundaries, thereby improving visual quality and prediction performance. It does this using adaptive filter taps to smooth the block boundaries and remove edge artifacts that are formed as a result of block coding. Unlike older standards like MPEG2 and MPEG4 Part 2, the in-loop filtering in H.264, H.265 and VP9 is part of the normative process. This means that it is applied in the encoding and decoding pipeline after decoding the pixels and before these pixels can be used for further prediction.

9.1   WHY IS IN-LOOP FILTERING NEEDED?

As we have seen in previous chapters, quantization is a lossy process that results in reconstructed pixels. These are approximations of the original, source pixels. The process of transforms and subsequent quantization happens on a block level, in block-based processing architectures. These tend to introduce artificial blocking artifacts along the edges of pixels. This can impair compression efficiency. As an example, assume we have the following two 4x4 neighboring blocks that are encoded:

150	151	153	155	157	159	161	163
150	151	153	155	157	159	161	163
150	151	153	155	157	159	161	163
150	151	153	155	157	159	161	163

The process of quantization and inverse quantization results in the following blocks.

152	152	152	152	160	160	160	160
152	152	152	152	160	160	160	160
152	152	152	152	160	160	160	160
152	152	152	152	160	160	160	160

While each of the 4x4 blocks that are separately transformed and quantized appear to be a reasonable approximation of the original input, the local averaging has resulted in a stark discontinuity along the 4x4 vertical edge that was not present in the source. This is quite common in block processing.

Further processing is needed to mitigate the impact of this artificially created edge. The deblocking process works to identify such edges, analyze their intensities and then applies filtering across such identified block boundaries to smooth off these discontinuities. The block boundaries thus filtered usually include the edges between transform blocks and also the edges between blocks of different modes.

In HEVC, the in-loop filtering is pipelined internally to two stages with a first stage deblocking filter followed immediately by a *sample adaptive offset* (SAO) filter. The deblocking filter, similarly to what occurs in H.264, is applied first. It operates on block boundaries to reduce the artifacts resulting from block-based transform and coding. Subsequently, the picture goes through the SAO filter. This filter does a sample-based smoothing by encompassing pixels that are not on block boundaries, in addition to those that are. These filters operate independently and can be activated and configured separately. In VP9, an enhanced in-loop filter with higher filter taps is used and SAO filter is not part of the standard. In the following sections, we shall explore both these filters in detail.

9.2   DEBLOCKING FILTER

This is the first filter used in H.265 and the only filter used in H.264 and VP9. It applies a boundary edge, pixel-adaptive smoothing operation across block edges. Filtering on picture boundaries is omitted. This filter provides the following two significant benefits.

1.   By selectively and adaptively smoothing out block edges, the decoded pictures look much better thereby enhancing the overall visual experience. This is especially true at low bit rates when block edge artifacts tend to become more visible due to higher quantization levels.
2.   As the filtering is applied after reconstructing the pixels but before these pixels can be used for prediction, the efficiency of the prediction is improved. This results in fewer and smaller residuals, hence better coding gains.

It should be noted that enabling the filter and determining the filter strengths are done adaptively based on the characteristics of the neighboring blocks. These characteristics include blocks coding modes, the QP values used and the boundary pixel differences. This ensures that an optimal level of filtering customized to the specific block boundary is applied, neither too little nor too much. It also ensures that, while edge artifacts are removed, sharp edges and details inherent in the picture are carefully analyzed and retained.

In HEVC, the deblocking filter can be disabled or enabled by signaling in the stream-level header called *picture parameter set* (PPS) and further parameters can be specified in the slice header. It is also disabled when a lossless coding mode is used, such as explicit PCM coding or in scenarios where transform and quantization is bypassed. In VP9, there is no sequence header signaling and deblocking parameters are included as part of the picture header. All these standards perform deblocking in two stages. First, the vertical edges are filtered, and then horizontal edges are processed on the vertically processed pixels.

As an example, Figure 78 illustrates the order in which the internal edges in a 64x64 superblock are filtered in VP9. The loop filter operates on a raster scan order of superblocks. For each superblock, it is first applied to vertical boundaries as shown in the thick lines. It is then applied to horizontal boundaries as shown by the dotted lines. The numbers in Figure 78 also indicate the order in which the filtering is applied.

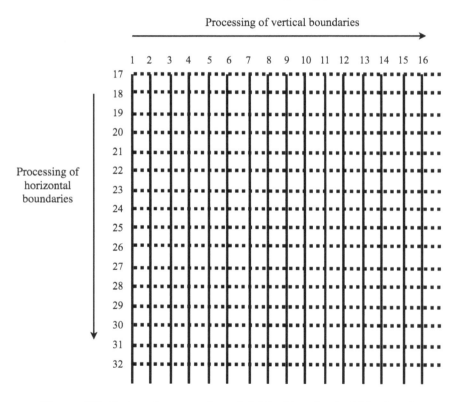

Figure 78: Order of processing of deblocking for 4x4 blocks of a superblock in VP9.

VP9 provides up to four deblocking filter options. These operate on up to 7 pixels on either side of the edges. HEVC provides two filters that modify up to 3 pixels on either side of the block boundary. While such details of filtering can vary across standards, conceptually they are identical and follow similar processing techniques that will be explained here.

9.2.1   DEBLOCKING PROCESS

The complete deblocking process involves the following three stages. The first two stages are analysis and decision steps. The last stage is the actual filtering step.

9.2.1.1   DETERMINING EDGES IN NEED OF DEBLOCKING

This is the first processing stage wherein the edges within the block are identified and marked for deblocking. Not all edges will be deblocked and this stage helps in marking which ones will be. If the standard prescribes that only edges of a certain sizes will be deblocked then other edges are marked for non-filtering. In H.265, deblocking is done only on 8x8 block boundaries whereas in VP9, it is done along 4x4, 8x8, 16x16 and 32x32 transform block boundaries. Similarly, under certain conditions, samples that are on the picture boundary or tile boundary pixels may be excluded from deblocking.

9.2.1.2   DERIVING THE FILTER STRENGTHS

Once the edges that need filtering are identified, the next step is to determine how much filtering is required for each edge by using the boundary strength parameter. This parameter is usually determined by inspecting a set of pixels on each side of the edge and also inspecting the characteristics of the edge, such as the prediction modes, motion vectors and transform coefficients used by the neighboring blocks.

The first idea here is to identify boundary areas with high probability of *blocking distortion* such as the boundary between intra coded blocks or blocks with coded transform coefficients. In such areas, stronger filtering will need to be applied.

Additionally, real edges around objects in the source also have significant pixel variance across the edge boundaries and will need to be preserved and not filtered. This is gauged by looking at the boundary pixel variance in conjunction with the corresponding block QP values. When the QP value is small, it is likely that boundary distortions due to block processing are low and therefore any significant gradient across the boundary is likely an edge in the source that needs to be preserved. Conversely, when the QP is large there is a strong possibility of blocking distortion that warrants a stronger filtering be applied.

9.2.1.3 APPLYING FILTERING

The final step is to actually apply the filtering along all horizontal and vertical edges that have been identified for filtering in the earlier steps. A corresponding calculation in this step is to decide how many pixels around the edges will be deblocked and adjusted. This is done based on the filter strength decision from stage 2.

If it's determined that a strong filtering is needed, more pixels are affected and vice versa. H.265 filters up to 3 pixels on either side of the edge for luma and one pixel on either side for chroma. VP9 has extended modes that filter up to 7 pixels per side for both luma and chroma.

9.2.2 FILTERING EXAMPLE

Figure 79: akiyo clip encoded at 100 kbps with deblocking.

The figures 79 and 80 illustrate the visual effects of deblocking through an example. These show the frame number 260 of the akiyo newsreader clip at CIF resolution. The clip is encoded using x265 H.265 encoder at a low bit rate of 100 kbps. This would have a high QP.

Figure 80: akiyo clip encoded at 100 kbps with deblocking disabled.
source: https://media.xiph.org/video/derf/

Figure 79 shows the encoded clip with deblocking filter enabled. Figure 80 illustrates the output with the filter disabled. While the deblocking filter does not necessarily soften all the blocking artifacts, it has worked to significantly improve the visual experience. This can be seen especially around the eyes and chin areas of the face and jaggies along the folds in the clothes. It should be noted that at low bit rates care should be taken to balance the filtering process and the preservation of real edges in the content.

9.3   SAO

The sample adaptive offset (SAO) filter is the second stage filter. It is used in-loop exclusively in H.265 after the deblocking filter is applied. This is illustrated in Figure 81, below. While the deblocking filter primarily operates on transform block edges to fix blocking artifacts, the SAO filter is used to remove ringing artifacts and to reduce the mean distortion between reconstructed and original pictures [1]. Thus, the two filters work in a complementary fashion to provide good cumulative visual benefit. In signal processing, low-pass filtering operations that involve truncating in the frequency domain cause ringing artifacts in the time domain. In modern codecs, the loss of high-frequency components resulting from finite block-based transforms results in ringing. As high frequency corresponds to sharp transitions, the ringing is particularly observable around edges. Another source of ringing artifact is the use of interpolation filters with a large number of taps that are used in H.265 and VP9.

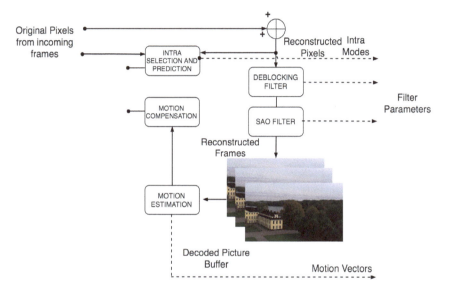

Figure 81: Video decoding pipeline with in-loop deblocking and SAO filters.

The SAO filter provides the fix by modifying the reconstructed pixels. It first divides the region into multiple categories. For each category, an offset is computed. The filter, then, conditionally adds the offset to each pixel within every category. It should be noted that the filter may use different offsets for every sample in a region. It depends on the sample classification. Also, the filter parameters can vary across regions. By doing this, it reduces the mean sample distortion of the identified region. The SAO filter offset values can also be generated by using any other criterion other than minimization of the regional mean sample distortion. Two SAO modes are specified in H.265, namely, *edge offset mode* and *band offset mode*. While the edge offset mode depends on directional information based on the current pixels and the neighboring pixels, the band offset mode operates without any dependency on the neighboring samples. Let us explore the concepts behind each of these two approaches.

9.3.1   EDGE OFFSET MODE

Figure 82: Four 1D patterns for edge offset SAO Filter in HEVC.

In H.265, the edge offset mode uses 4 directional modes (using neighboring pixels) as shown in Figure 82 (Fu, et al., 2012) [1]. One of these directions is chosen for every CTU by the encoder using rate distortion optimization.

For every mode, the pixels in the CTU are analyzed to see if they belong to one of the four categories, namely, 1) local valley, 2) concave corner, 3) local peak or 4) convex corner. This is shown in Figure 83.

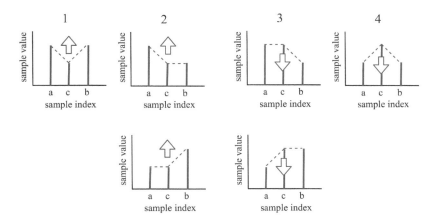

Figure 83: Pixel categorization to identify local valley, peak, concave or convex corners [1].

As categories (1) and (2) have the current pixels in a local minimum compared to their neighboring pixels, positive offsets for these categories would smooth out the local minima. A negative offset would in turn work to sharpen the minima. The effects, on the other hand, would be reversed for categories (3) and (4) where negative offsets result in smoothing and positive offsets result in sharpening.

9.3.2   BAND OFFSET MODE

In band offset (BO) mode, the sample value range is equally divided into 32 bands (for 8-bit pixels), each with eight values. For every band, the BO that is the average difference between the original and reconstructed samples is calculated and sent in the bitstream. To reduce complexity and signaling, in HEVC, not all band offsets are signaled. Instead, only four offsets corresponding to bands with pixel values between 72 to 104 are signaled. If a sample belongs to any other band, BO is not applied. In Figure 84 [1], the horizontal axis denotes the sample position and the vertical axis denotes the sample value. The dotted curve is the original samples. The solid curve is the reconstructed samples. As we see in this example, for these four bands, the reconstructed samples are shifted slightly to the left of the original samples. This results in negative errors that can be corrected by signaling positive BOs for these four bands.

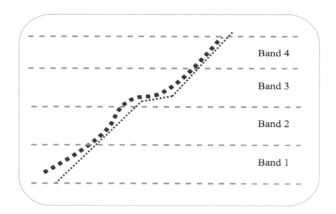

Figure 84: Illustration of BO in HEVC, where the dotted curve is the original samples and the solid curve is the reconstructed samples.

9.3.3   SAO IMPLEMENTATION

The implementation of SAO can be done using rate-distortion optimization techniques. A range of offset values is carefully selected and added to the pre-SAO pixels. The distortion between the source samples and post-SAO reconstructed samples is then calculated. This is done for all selected offsets across all bands or EO classes. The best offset and edge offset class for EO mode and the best offset for each band, in case of BO, are then obtained by choosing the ones that minimize the RD cost. The delta cost for the entire CTU can then be calculated by computing the difference between the best post-SAO cost and the pre-SAO cost. If this delta cost is negative, then the SAO filter is enabled for the CTU.

9.4 SUMMARY

- A deblocking filter is applied to the reconstructed pixels to remove the block artifacts around block edge boundaries, thereby providing the dual advantage of improving visual quality and prediction performance.
- In HEVC, the in-loop filtering is pipelined internally to two stages with a first stage deblocking filter followed by a SAO filter.
- Deblocking filter implementations vary in the number of pixels they operate and improve on. VP9 provides up to four deblocking filter options that operate on up to 7 pixels on either side of the edges. HEVC provides two filters that modify up to 3 pixels on either side of the block boundary.
- Two SAO modes are specified in H.265, namely, edge offset mode (EO) and band offset mode (BO).
- EO mode depends on directional information derived from the current pixels and the neighboring pixels. BO mode operates without any dependency on the neighboring samples.

9.5 NOTES

1. Fu C, Alshina E, Alshin A, et al. Sample adaptive offset in the HEVC standard. IEEE Trans Circuits Syst Video Technol. 2012;22(12):1755-1764.
    http://citeseerx.ist.psu.edu/viewdoc/download?doi=10.1.1.352.2725&rep=rep1&type=pdf. Accessed September 22, 2018.

10 MODE DECISION AND RATE CONTROL

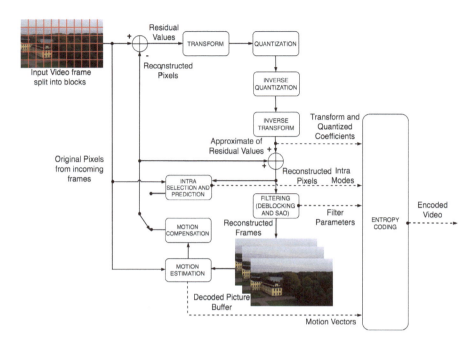

Bitrate is the amount of data in bits that is used to encode one second of video. It is usually expressed in megabits per second (Mbps) or kilobits per second (kbps). Bitrate is a critical parameter of the video and affects the file size and overall quality of the video. In general, the higher the bitrate, the more bits there are available to encode the video. This means better video quality but usually comes at the expense of bigger file size. When the application provides a target bitrate to the encoder, the job of the encoder is to allocate the available bits intelligently across the video sequence, keeping track of the video complexity and delivering the best possible picture quality.

Internally, the video encoding process is highly complex because it involves a combination of computationally intensive mathematical operations together with the need to take complex decisions at various stages. As we know, video encoding is a lossy process and the loss is introduced primarily by the quantization process. At the outset, how much loss is required, or, in other words, how much compression ratio is needed, is determined by the overall encoding settings and parameters. The settings include channel capacity available and encoding mode required. The parameters include output bit rates and latency settings. These are the constraints under which the encoder operates. However, the actual encoding process is much more complex, in that the encoder must take decisions on all levels, starting from a GOP all the way down to superblocks and sub-blocks. By understanding picture complexity, the encoder has to determine the optimal picture types, motion vectors and prediction modes for every block in every picture in the video sequence. This complex process is called *mode decision*. Every block then also has to be encoded using an optimal number of bits. The process of assignment of bits across different parts of the sequence is called *rate control*.

In the first section of this chapter, we will deal with the process of mode decision. The latter half of the chapter will cover topics in rate control. We will also see in this chapter how these processes are intertwined with one another.

10.1   CONSTRAINTS

Given specific settings, including bitrate, latency and so on, the fundamental challenge for any encoder is how to optimize the output-encoded picture quality such that it can either:

a) maximize the output video quality for the given bit rate constraints, or
b) minimize the bit rate for a given output video quality.

While doing the above, the encoder must also ensure that it operates within its several constraints. Some of these are outlined below:

Bitrate. The encoder has to ensure it produces an average bitrate per this setting. Additional constraints may also be imposed, such that it may also be required to operate within a set of maximum and minimum bitrate limits. This is especially the case in constant bitrate mode where, usually, the channel capacity is fixed.

Latency. Latency is defined as the total time consumed between the picture being input to the encoder and being output from the decoder and available for display. This interval depends on factors like the number of encoding pipeline stages, the number of buffers at various stages in the encoder pipeline and how the encoder processes various picture types, such as B-pictures, with its internal buffering mechanisms. The interval also includes the corresponding operations from the decoder. The term, latency, usually refers to the combined latency of both the encoder and the decoder.

Buffer Space. When the decoder receives the encoded bitstream, it stores it in a buffer. There, the decoder smooths out the variations in the bitrate so as to provide decoded output at a constant time interval. Conversely, this buffer also defines the flexibility that the encoder has, in terms of the variability of its bitrate, both instantaneously and at any defined interval of time.

The buffer fullness at any time is thus a difference between the bits encoded and a constant rate of removal from the buffer that corresponds to the target bitrate. The lower boundary of the buffer is zero and the upper boundary is the buffer capacity. H.264 and H.265 define a *hypothetical reference decoder* (HRD) buffer model. This model is used to simulate the fullness of the decoder buffer. This aids the rate control in producing a compliant bitstream.

The encoder thus has to tightly regulate the number of bits sent in any period of time. This is to ensure that the decoder buffers are never full or empty. This is especially true for hardware decoders that often have limited memory buffers. When the decoder buffer is full, no further bits can be accommodated, and incoming bits may be dropped. On the other hand, if the decoder has consumed all the bits and the buffer becomes empty, it may not have anything to display except the last decoded picture. This may manifest as undesirable pauses in the output video.

Encoding Speed. Typically, encoding applications get classified as either real-time or non real-time encoding and most encoders are designed for one or the other. Examples of real-time encoding applications include live event broadcasting. Here, the camera feed reaches the studios where it's processed, encoded and streamed over satellite, cable or the internet in real time. In real-time encoding, if the output frame rate is 60fps, then the encoder has to ensure it can produce an encoded output of 60 frames in every second of its operation.

Non-real time encoding, or, offline encoding, has the luxury of time to perform additional processing in an effort to improve the encoding quality. A typical example is video on demand streaming where the encoding is done on all the content offline and stored in servers and the requested video is fetched, streamed and played back upon demand.

Operating within these constraints, the encoder has to take decisions at all stages, including selecting picture types, selection of partition types for the coding blocks, selection of prediction modes, motion vectors and the corresponding reference pictures they point to, filtering modes, transform sizes and modes, quantization parameters, and so on. By taking these decisions at various stages, the encoder strives to optimize the bit spend both within and across various pictures to provide the best output picture quality.

This video quality is measured objectively by comparing the reconstructed video (encoded and decoded) to the original input video using a mathematical formula called *distortion measure* that is usually computed pixel-by-pixel and averaged for the frame. As the distortion measure is an indication of how much different the reconstructed frame or block is from the original, the higher this number the worse the quality associated with the selected block and vice versa.

10.2   DISTORTION MEASURES

In this section, we will explore two widely used distortion measures, namely, the *sum of absolute differences* (SAD) and *sum of absolute transform differences* (SATD). In these methods, a pixel-by-pixel operation is performed and all pixels in both the original and the reconstructed picture are used. As these measures are used for decisions at every block level, the computations are usually performed in the encoder at a block level.

10.2.1   SUM OF ABSOLUTE DIFFERENCES

To arrive at this metric, the absolute differences between each pixel in the original picture block and the corresponding pixels in the reconstructed block are calculated. The sum of these differences is the SAD for the block. In signal processing, this corresponds to the *L1 norm* and is expressed mathematically by the following equation:

$$\text{SAD} = \sum_n \sum_m |\, c_{n,m} - r_{n,m}\,|$$

where $c_{n,m}$ corresponds to the current picture block samples and $r_{n,m}$ corresponds to the reconstructed block samples. The SAD can be used in motion estimation to compare the similarities of the current block of pixels to the block being pointed to by the motion vector in the reconstructed picture set used for prediction.

10.2.2   SATD (SUM OF ABSOLUTE TRANSFORM DIFFERENCES)

This is a metric that has been used since H.264 where Hadamard transforms were introduced for transform of residual samples. The SATD is similar to the SAD except that it does an additional step of computing the Hadamard transform of the residual samples and then calculates the sum of absolute values of the Hadamard-transformed residuals. By incorporating the integer Hadamard transform in the cost computations, this metric is better able to portray the actual cost of coding the resulting residual values when Hadamard transforms are used. In such scenarios, this measure results in better decisions in the motion estimation process.

$$\mathbf{T} = \mathbf{H} \,.\, (\mathbf{C} - \mathbf{R}) \,.\, \mathbf{H}^{\mathrm{T}}$$

$$\text{SATD} = \sum_n \sum_m |\, t_{n,m}\,|$$

where $\mathbf{C}$ is the matrix corresponding to the current picture block samples and $\mathbf{R}$ is the matrix representation of the reconstructed block samples. $\mathbf{H}$ is the Hadamard transform matrix and $\mathbf{T}$ is thus the result of Hadamard transform of the residual samples whose sum of absolute values results in the SATD metric.

10.3   FORMULATION OF THE ENCODING PROBLEM

Thus, the fundamental encoding problem is an optimization problem that can be stated as minimization of the distortion between the input video and its output reconstructed video, subject to a set of constraints including bitrates and coding delay, as we have seen already.

Given the large number of encoding parameters, the above minimization problem is broken down into smaller minimization problems.

As mentioned earlier, the encoder has to decide, for each superblock or CTU, what block partitioning to use, what coding mode to use and what prediction parameters to choose. This has to be done for every block in every picture in the video sequence by keeping in mind at every step, the output bit rate and distortion produced as a result of the previous selections. Thus, the task for the encoder is to find the best coding modes for every picture, such that the selected distortion measure (D) is minimized while always subject to the rate constraint (R_c).

Mathematically, this can be expressed as:

$$\text{min } D \text{ with } R < R_c$$

Let us assume we break down the video into n blocks. Let P represent the set of the n coding decisions to make on these 'n' blocks. These coding decisions directly affect the distortion (D) and rate of bits produced (R). Hence, we can say 'D' and 'R' are functions of P. The *constrained* minimization problem can be thus represented as:

$$\text{min } D \text{ (P) with } R \text{ (P)} < R_c$$

One way to solve this problem is to convert this *constrained* minimization problem into an *unconstrained* minimization problem using the method of Lagrange multipliers. In this method, the constraint function is appended to the function to be minimized by multiplying it with a scalar called *Lagrange multiplier*. This becomes a Lagrangian function (say J) and the solution to the original *constrained* problem is obtained by solving for both an optimal P_n and an optimal set of Lagrange multipliers (say λ). This is mathematically represented as:

$$J \text{ (P, } \lambda) = D \text{ (P)} + \lambda \cdot R \text{ (P)}$$

$$J_{opt} = \text{min } J \text{ (P, } \lambda)$$

To simplify the implementation, it's assumed that the coding decisions of the n blocks are independent of each other and thus the joint cost J can be derived as the sum of joint costs of the n blocks.

$$J \text{ (P, } \lambda) = \sum_n J_n \text{ (}P_n, \lambda)$$

The optimal solution of this optimization problem J (P, λ) can be obtained by independently selecting the coding modes P_n for the n blocks. Under this simplification, the combined optimal cost J_{opt} is the sum of optimal costs of the n blocks.

$$J_{opt} = \sum_n \min J_n (P_n, \lambda)$$

The encoder has to estimate the number of bits that will be spent if a particular mode were selected to encode the block and then compute the cost of using that mode. By iterating across all the modes and computing the corresponding bit costs, the encoder can analyze and select the mode best able to minimize the cost of encoding. The above optimization technique has been widely deployed, thanks to its effectiveness and simplicity. Using this approach, the computational time spent evaluating all the modes and testing their costs for minimality has a significant performance impact. The decisions made in one block tend to have cascading effects on decisions made in other blocks, both spatially and temporally, and the design of video encoders involve simplifications which usually ignore these effects.

10.4   RATE DISTORTION OPTIMIZATION

The overall bit rate of the encoded stream depends on the modes selected by the encoder for the blocks, the motion vectors and the bits used to encode the transformed residuals after quantization. The quantized coefficients account for typically 60-70% or more bits in the bitstream, hence are the significant focus of rate control. By increasing the Q_p, the bit rate (R) is reduced and vice versa. However, this comes at the expense of the video quality of the resulting bitstream. Increased Q_p typically results in greater distortion (D) and hence lower video quality.

It is thus seen that the Q_p directly influences the balance between distortion (D) and rate (R). Therefore, control over the Lagrangian parameter λ (which defines the relative weights of R and D) is synonymous with control over the Q_p. The Lagrangian parameter λ is thus built into the rate control process to avoid loss in coding performance while effectively maintaining the required bitrate. The goal of the rate control algorithm is to accurately achieve the target bitrate with the lowest distortion.

Figure 85 shows a typical R-D curve of a video where point A is on the R-D curve, whose bitrate is R_A. As seen from this figure, when the bitrate increases the distortion decreases and vice versa. In an R-D framework, the goal of rate control is to derive the best operating point around the target bitrate R and the challenge for the rate control algorithm is to estimate the R-D function. There could be several ways to approximate this. The traditional approach is to establish a mathematical relationship between bitrate R and Q_P and perform the R-D optimization in the Q_P domain.

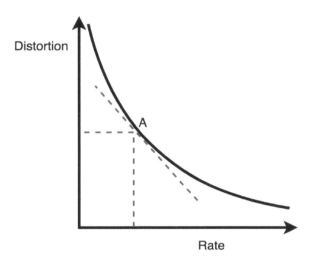

Figure 85: Rate distortion curve.

It should be noted that the rate distortion optimization technique is applied at least across the following three operations in the encoder: 1) motion estimation in selecting the best possible prediction, 2) mode decision while selecting the best possible block modes and partitions, and 3) rate control to allocate the bits within and across the pictures. Of these operations, Q_P is directly associated only with the rate control operation wherein it directly controls the bitrate. A direct relationship between non-residual bits like motion vectors, prediction modes, and so on does not exist with Q_P. It is not applied to these parameters. Therefore, using Q_P alone to model the rate potentially can result in inaccurate costs estimates for these processes.

However, the one parameter that's universal across all stages where rate distortion optimization is used is λ and a better estimate of bits could be performed by exploiting the robust relationship between λ and R. This, in turn, could provide a more precise bits estimate for every picture.

10.5   RATE CONTROL CONCEPTS

As we know, the goal of rate control is to accurately achieve the target bitrate with the lowest distortion. The rate control algorithm maintains a bit budget. It allocates bits to every picture and every block within each picture by analyzing the video complexity and keeping track of previously allocated bits and the target bitrate. Rate control is not a normative part of the coding standard but is a critical feature distinguishing one encoder from another.

As the picture complexity changes in the video sequence, the encoder keeps track and dynamically changes parameters like QP to accurately follow the requested bitrate while maintaining the picture quality. The heart of the rate control algorithm is a quantitative model that establishes a relationship between the Lagrangian parameter λ, Q_P, the bitrate and a distortion measure.

Rate control algorithms contain two important functions or sections. The first function is bit allocation. This considers key requirements and allocates bits, starting with GOP-level bits allocation and narrowing down in granularity to a basic unit-level bit allocation. The second function concerns how the target bitrate is achieved for a specific unit. This uses models like the rate distortion optimization models we discussed earlier. In this section, we will delve into these two functions in greater detail. We will explore how they are intertwined and work together in helping to achieve a bitrate that's close to the target bitrate.

10.5.1   BIT ALLOCATION

This section illustrates the concepts behind how the rate control algorithms allocate bits at various levels of encoding through a simple mechanism. In this scheme, bit allocation is done at the following three levels.

1.   GOP-level bit allocation
2.   Picture-level bit allocation
3.   Basic unit-level bit allocation

It should be noted that the first picture is usually treated in a special way, as the encoder usually has no a priori knowledge of the encoding content and it is impossible for an encoder to accurately estimate the number of bits for the first picture. However, this is usually solved by one or a combination of the following approaches.

1. **Preprocessing.** A preprocessing step like *lookahead* does a first-pass encoding to provide estimates of crude complexity and encoding parameters.
2. **User Input.** Use of user-specified encoding parameters like initial QP for the first picture means that the encoder doesn't need to calculate the target bits for this picture. However, this is not a preferred method.
3. **Approximate Bits Estimation.** A better approach over a blind initial Q_P can be to estimate bits based on a set criterion like target bit rate. For example, one simple way is to just allocate about five to six times the average bits per picture based on the target bitrate. This allocation would occur as the first picture is intra coded.

While different strategies can result in either successful or failed prediction of the number of bits needed in a short, initial time frame, it is important to note that the performance of rate control should be measured over a longer time period. The algorithm has better opportunities to adapt and adjust the bitrate over the longer term to make up for any inaccurate initial bits estimation.

10.5.1.1 GOP LEVEL BIT ALLOCATION

This is the top-level bit allocation. A target bitrate for the entire GOP is calculated based on the target bit rate and the fullness of the decoder buffer. As the picture complexity can vary dynamically across various scenes, hence across GOPs, it's difficult to assign a fixed target for bits to every GOP. A better approach is to offset the bits target of the current GOP by how many more or fewer bits the previous GOPs used, compared to the target bitrate.

If the previous GOPs used more or fewer bits, the current GOP should correspondingly cost fewer or more bits. Enhanced strategies also use a sliding window of pictures of about 30-60 frames in which adjustments are made to make the bitrate, hence the video quality adjustment, smoother. The size of the sliding window is bigger than the GOP size and a larger window size provides more room leading to smoother adjustments. Once the GOP target bitrate is determined, it is fed down the layers for accurate picture bit allocation and basic unit bit allocation.

10.5.1.2 PICTURE LEVEL BIT ALLOCATION

Bits that are left available in the GOP are assigned by the picture-level bit allocation algorithm to the pictures, either uniformly or in accord with an assigned picture weight. One method that has been explored is hierarchical bit allocation. In this method, different pictures belong to one of several predetermined levels and every picture is assigned bits in accordance with its level weight. Hierarchical bit allocation has been found to achieve performance improvements because it aligns the bit allocation with other coding parameters of the pictures.

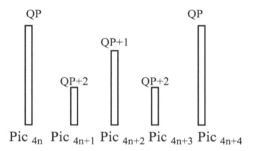

Figure 86: Hierarchical picture level bit allocation scheme.

Figure 86 shows a typical hierarchical bit allocation scheme available in HEVC software. In this scheme there are three levels assigned in a hierarchical pattern. Pictures Pic_{4n} and $Pic_{4(n+1)}$ belong to the first level. Level 1 has a picture-level QP value that results in the most bits. Pictures Pic_{4n+1} and Pic_{4n+3} belong to the third level. Level 3's QP has been increased the most, resulting in the fewest bits. Picture Pic_{4n+2} is belongs to the second level. It has a mid-level QP adjustment and therefore a bit consumption between the other two levels.

10.5.1.3 BASIC UNIT LEVEL BIT ALLOCATION

With this approach, a scalable rate control to different levels of granularity is possible within each picture. The level of granularity could be a slice, a block, or any contiguous set of blocks. This granular level is called a basic unit (BU) of rate control for which usually distinct values of QP are used. The BU level bit allocation algorithm is quite similar to the picture-level bit allocation algorithm. It allocates the leftover bits to the remaining BUs in accord with the weight of the BU. The weights can be predetermined or also may be calculated dynamically. The latter may employ a complexity metric such as the estimated *mean average difference* (MAD) of the prediction error of the collocated BU in the previously coded picture belonging to the same picture level. Ideally, the MAD would be calculated after encoding the current picture. However, that would require us to encode the picture again after the corresponding QP is selected. Instead, we assume that this complexity metric varies gradually across pictures and we use an approximation from the previous pictures belonging to the same level.

10.5.2   RDO IN RATE CONTROL

In the previous section, we explained how bits are allocated at both the picture and the BU level. Using the target rate information and a corresponding complexity (distortion) measure, the encoder has to now compute the λ value used for encoding.

10.5.2.1 DETERMINATION OF λ

The rate-distortion model, as we explained earlier, does just this by using a mathematical model to derive the λ value from the target bitrate for a picture or BU. It should be noted that the initial values used by the model are not fixed. Different sequences may have quite different modeling values and these values will also get updated as the encoding progresses. Thus, the model adapts dynamically. After encoding one BU or picture, the actual encoded bitrate is used to update its internal parameters and correspondingly derive future updated λ values.

10.5.2.2 RDO CODING

Coding standards like H.264, H.265 and VP9 provide several partition sizes and inter and intra prediction modes, including direct or skip modes. A critical function in encoding is to ensure that the correct modes are selected. This is crucial in unlocking significant bit rate reductions. However, going through all the different combinations and coming up with optimal decisions comes at the expense of increased computational complexity. One way to select the optimal modes is to use the rate-distortion optimization (RDO) mechanism that was described in an earlier section. This uses the λ value, the target bitrate, and distortion metrics to do the following:

1) perform an exhaustive calculation of all modes to determine the bits used and corresponding distortion of each mode,
2) use its internal model to compute a metric that takes as input the bitrate and distortion calculated for every mode, and
3) select the mode that minimizes the metric.

Thus, once the λ value is determined for a picture or BU, all the coding parameters including partition and prediction modes and motion vectors can be determined by using exhaustive RDO search. The QP value could be also determined by an exhaustive QP optimization to achieve the optimal RD performance or it could be simplified and derived using a model.

It should be noted that the RDO process is complementary to the rate control process as it does not directly control the QP value. The interplay of RDO and rate control is sometimes seen as a chicken and egg problem, because RDO in effect influences the rate control algorithm. As we saw earlier, the distortion measure MAD is needed by the rate control BU-level bit allocation algorithm. However, it's not accurately available until all the prediction modes and thereby residuals are computed using RDO. This is why the rate control algorithm uses an estimate for MAD from previously encoded pictures.

Thus, these two processes are decoupled using approximations as needed to keep the solution computationally feasible. To maintain a consistent video quality and experience, both λ and QP should not change drastically. They are then passed through a QP limiter step that limits their variations both at a picture and a BU level.

10.5.3 SUMMARY OF RATE CONTROL MECHANISM

Figure 87 is adapted from the whitepaper titled Rate control and H.264 [1]. It provides an overall summary of the rate control mechanism that has been discussed in previous sections. The inputs to this process are:

  a) the target bitrate that is usually provided by the application or through user input,

b) buffer capacity, and

c) initial buffer occupancy condition.

At the start of the encoding process, the demanded bitrate input is fed to the virtual buffer model (if present). This then provides information on buffer fullness.

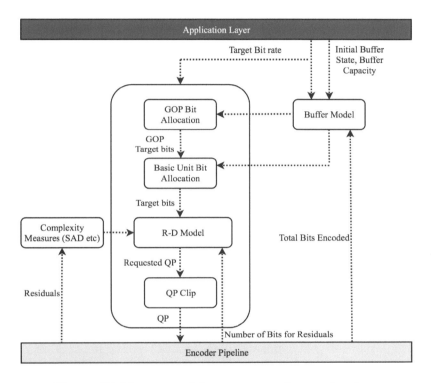

Figure 87: Elements of the rate controller mechanism.

During the process of encoding, buffer fullness is always kept track of by monitoring the total bits that have been encoded thus far and the rate of removal of the bits from the buffer. The buffer fullness information, along with the target bitrate, are fed as input to the bit allocation units. These are then used to compute the GOP level, picture level and basic unit-level target bits.

The BU-level target bits, along with a surrogate for spatial picture complexity information like MAD (usually stored from previous pictures), are then used as the inputs for the rate and distortion calculations in the R-D model. The R-D model also takes as an input an initial QP value. This will have been computed based on the target bitrate and updated according to the running bitrate. The output of the R-D model is a corresponding target QP value to encode the BU. This target QP value then passes through a QP limiter block that analyzes the QP values over previous QP values. This ensures that a smooth transition is provided and any dramatic changes in the QP values are smoothed out. The output of the QP limiter block is the final target QP for the BU.

When the BU is quantized and encoded, the following parameters are fed back into the model for future computations.

a) Total bits encoded for the BU
b) Residual bits encoded for the BU
c) Actual residual values

The total bits parameter is used to update the buffer fullness in the virtual buffer. The residual bits parameter is updated to provide accurate rate information to the RD model. The actual prediction residuals are fed back into the complexity estimator (MAD).

This framework allocates QPs and corresponding bits to different picture types flexibly using the target picture allocation mechanism at the picture level.

10.6 ADAPTIVE QUANTIZATION (AQ)

It's also possible to get significant QP variations across different BUs in the same picture, in accord with their complexity variations. This tuning is called *adaptive quantization (AQ)* and is more a visual quality fine tuning tool than a rate control mechanism. Modern encoder implementations have built-in mechanisms to analyze the incoming pictures before they are encoded. These pre-analysis tools, which are also called *lookahead*, often perform spatial and temporal scene content analysis.

By using objective metrics, they provide scene complexity information. AQ algorithms make use of this scene complexity information to optimally allocate the bits across different BUs to provide immense visual quality benefits. As we know, our eyes are more sensitive to flatter areas in the scene and are less sensitive to areas with fine details and higher textures. AQ algorithms leverage this to increase the quant offset in higher textured areas and decrease it in flatter areas. Thus, more bits are given to areas where the eyes are sensitive to visual quality impacts.

This is illustrated in Figure 88 where higher textured areas like the text around the corners are given a positive quant offset and are indicated with a lighter shade. Other flatter areas are given more bits with a negative quant offset and are shown in a darker shade. Thus, AQ serves as a good visual quality fine tuning tool that allows a balanced spatial quality throughout the picture.

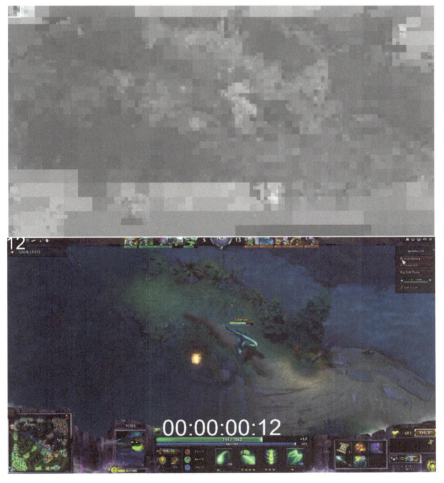

Figure 88: Heat map showing quant offset variation using Adaptive Quantization [2].

10.7   SUMMARY

- Bitrate is a critical parameter that affects the file size and overall video quality. At higher bitrates, more bits are available to encode the video. This results in better video quality but comes at the expense of bigger file size.
- Typical constraints that the encoder has to operate with are latency, bitrate, buffer space, and encoding speed.
- Given a specific setting that includes bitrate, latency, and so on, the fundamental challenge for any encoder is how to optimize the bitrate and output encoded picture quality. The encoder has to either maximize the output video quality for the given bitrate or minimize the bitrate for a set video quality.
- Video quality is quantified mathematically using a distortion measure. This is usually computed at every pixel and averaged for the frame. This provides a good measure of similarity between the blocks that are compared.
- Encoding is an optimization problem where the distortion between the input video and its output, reconstructed video is minimized, subject to a set of constraints including bitrates and coding delay.
- Rate control algorithms maintain a bit budget and allocate bits to every picture and every block within each picture by analyzing the video complexity and keeping track of previously allocated bits and the target bitrate.
- Rate control algorithms contain two important functions: 1) determining and allocating target bits and 2) achieving target bit rate.

10.8   NOTES

1. Rate control and H.264. PixelTools Experts in MPEG. http://www.pixeltools.com/rate_control_paper.html. Published 2017. Accessed September 22, 2018.
2. DOTA2. xiph.org. Xiph.org Video Test Media [derf's collection]. https://media.xiph.org/video/derf/. Accessed Oct 30, 2018.

PART III

11 ENCODING MODES

In previous chapters we have explored at length how the encoder internally operates to allocate the target bits across the video sequence. Let us now review three important, application-level encoding modes and the mechanisms for bitrate allocation in each of these modes. These encoding modes are agnostic with respect to encoding standards. This means that every encoder can be integrated with one or more of the rate control mechanisms. All these modes are accessible in publicly available x264, x265, and libvpx versions of H.264, H.265 and VP9, respectively.

11.1 VBR ENCODING

Variable bit rate (VBR) is a mode wherein the encoder varies the amount of output bits per time segment, usually in seconds. In VBR mode, the encoder allows more bits to be allocated, as necessary, to the more complex segments of the video (like action or high-motion scenes) and uses fewer bits to encode simpler and more static segments. This means the encoder tolerates the dramatic fluctuations in bitrate needed without imposing severe restrictions. By doing optimal bit allocations as needed to encode the scene, the encoder is thus able to maintain the average bitrate and ensure the best video quality at the same time. For example, when we encode a video at 4 Mbps VBR, the encoder will vary the bit rate by giving some sections of frames as much as 6 or 7 Mbps while giving others only 2 or 3 Mbps. Eventually, however, the overall average rate across the whole stream or file would be 4 Mbps. The advantage of using VBR is that it produces a better-quality video as the encoder operates with less rigid constraints. The disadvantages are that it may consume more bits and result in poorer adherence to the target bitrate especially if the stream has lots of complex scenes. Unrestricted VBR without bitrate caps will result in packet drops when the instantaneous bitrate exceeds the channel bitrate. However, this can be avoided by specifying and imposing an upper limit on the instantaneous bitrate in what is called a *capped VBR* mode.

11.2 CBR ENCODING

In *constant bitrate* (CBR) encoding, on the other hand, the encoder closely tracks the bits usage. It imposes more rigorous constraints on the bitrate around periodic intervals. Also, it encodes the video at a more or less consistent bitrate by disallowing drastic bitrate swings (peaks or troughs) for the duration of the sequence. Variation exists among different frame types because these will have different data rates (e.g., I frames consume the highest number of bits followed by P and B Frames). However, the allocation of bits across a time segment will be closely monitored and bit expense averaged across the time segment. The encoder has to also consider the buffer model that the video playback devices will employ and closely adhere to this decoder buffer model such that, at any time during the stream, the buffers are neither full nor empty. This is required for a smooth playback.

In this mode, the encoder keeps track of the number of bits consumed and the available bits over a predefined buffer interval. It imposes bit constraints such that the bitrate is maintained over the buffer interval. The variations that happen are much smaller than in VBR. Encoding is constant over an interval, typically around 1 or 2s. The disadvantage with CBR is that, when there is increased activity in a scene that results in a bit rate demand higher than the target rate, the encoder has to prioritize adherence to the bitrate over quality and impose restrictions to keep the bit rate under check. This could potentially result in a lower picture quality relative to video encoded in VBR mode.

For example, when we encode a video at 4 Mbps CBR, the encoder will vary the bit rate by giving some frames as much as 6 or 7 Mbps while giving others only 2 or 3 Mbps but ensure that the bitrate does not exceed 4 Mbps in any given time period, say, 1s or 2s. Also, if there's a frame that requires 5 MB, the encoder may not permit it, depending on how many bits are available that can be expended for that frame. As you can imagine, if there are many frames that need more than 4 Mbps to encode, the output of CBR will look worse than that of VBR.

11.3   CRF ENCODING

The constant rate factor (CRF) mode is a new mode that has garnered a lot of attention in recent years. It is used as the default mode in many modern codecs and is a constant quality encoding mode, in that it prioritizes the quality metric and ensures a constant quality across all sections of the video. The traditional approach to achieving a constant quality was to use a constant QP (fixed QP) encoding mode where a fixed QP is applied to every picture, thereby compressing the pictures equally and resulting in a uniform quality across the sequence. For example, a fixed QP encoding with QP set to 25 will assign every frame the same QP value of 25. However, as this mode ignores any bitrate constraints, it typically results in large swings in the video bitrate across the sequence.

The CRF mechanism improves on this idea to maintain the required level of perceptual quality. One of the ways it does so is by leveraging the HVS and using motion as a metric in varying the QP across different pictures in the video. As we know, the human eye is more sensitive to changes in static uniform scenes and still objects and less sensitive to objects in motion. By using this HVS characteristic, the CRF bitrate algorithm is able to increase the QP and thereby reduce the bitrate accordingly in motion areas while increasing or maintaining the QP to provide more bits in areas of less motion. In areas of high motion, there is a lot going on in the scene for the eyes to perceive, resulting in not enough time to notice the slightly higher compression. However, in static areas, there isn't much happening and any minor change that affects this setting will be quickly perceived by the eye. Thus, in this mode, the encoder adjusts the QP to deliver a fixed perceptual quality output. For example, a CRF encoding QP = 25 will vary the QP, increasing it to, say, 28 for scenes with high motion and lowering it to, say, 23 for scenes with more static content.

While this approach is counterintuitive, it serves to improve subjective or perceived visual quality significantly. However, it should be noted that this mechanism could also result in a lower video quality, as objectively measured by PSNR. As perceptual quality also depends on uniformity in the pictures, the QP adjustments are always gradual. Also, using an increased QP might still result in a higher bit usage in the high motion areas just because of their sheer complexity.

11.4   WHEN TO USE VBR OR CBR?

When it comes to selecting VBR or CBR, it usually depends on the application in question. By default, it is always recommended to use some form of VBR encoding whenever the application permits it. However, in applications where the transmission infrastructure has a fixed bandwidth pipe, CBR is usually the preferred option as it provides much more reliability by ensuring the bitrate does not vary to overflow the available bandwidth pipe.

Also, some playback devices only support CBR mode. This is in order to limit the amount of data the decoder must internally buffer during the process of decoding and playback. For maximum device compatibility, CBR could provide a safer option. The common differences between CBR and VBR modes of operation are highlighted in Table 14.

Table 14: Comparison of bit allocations in CBR and VBR modes.

Constant Bit Rate (CBR)	Variable Bit Rate (VBR)
Variable video quality, usually worse than VBR	Constant, definable video quality and highest video quality
Predictable file sizes	Unpredictable file sizes
Compatible with most systems	Unpredictable compatibility
Transmission applications with fixed bandwidth pipe	Storage applications where only final size limit is defined

The following example illustrates how the VP9 reference encoder allocates bits in a sample CBR and VBR configuration for a test clip encoded at a target of 3 Mbps.

CBR Configuration:

```
./vpxenc test_1920x1080_25.yuv -o test_1920x1080_25_vp9_cbr.webm --codec=vp9 --i420 -
w 1920 -h 1080 -p 1 -t 4 --cpu-used=4 --end-usage=cbr --target-bitrate=3000 --
fps=25000/1001 --undershoot-pct=95 --buf-sz=18000 --buf-initial-sz=12000 --buf-optimal-
sz=15000 -v --kf-max-dist=999999 --min-q=4 --max-q=56
```

VBR 2-Pass Configuration:

```
./vpxenc test_1920x1080_25.yuv -o test_1920x1080_25_vp9_vbr.webm --codec=vp9 --i420
-w 1920 -h 1080 -p 2 -t 4 --best --target-bitrate=3000 --end-usage=vbr --auto-alt-ref=1 --
fps=25000/1001 -v --minsection-pct=5 --maxsection-pct=800 --lag-in-frames=16 --kf-min-
dist=0 --kf-max-dist=360 --static-thresh=0 --drop-frame=0 --min-q=0 --max-q=60
```

The graph in Figure 89 shows the trendline for the variation of bits over the duration of the clip. As we see from the figure, the bitrate fluctuations are far lower in the CBR mode compared to VBR mode. The test scripts that were used are as follows:

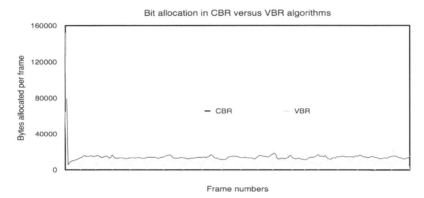

Figure 89: Comparison of bit allocations in CBR and VBR modes.

The following section highlights different typical encoding application scenarios and the choice of the rate control mode for each of these.

11.4.1 LIVE VIDEO BROADCASTING

This is encoding for video distribution over terrestrial, satellite or cable networks where the requirement is to encode and pack as many channels as possible in a fixed available bandwidth in real time. Video quality is of utmost importance here and is usually achieved at the expense of large latency and buffers, typically a few seconds. Usually, CBR or some form of statistically multiplexed VBR mode is used to encode video in this case. Statistical multiplexing is a VBR technique that leverages the relative complexity of all channels in a bit rate pool at any time and uses time division multiplexing algorithms to allocate bits to each of these channels in accord to their complexity. The overall channel pool bitrate remains CBR but within the pool, the individual channels themselves are allocated VBR dynamically.

11.4.2 LIVE INTERNET VIDEO STREAMING

This use case includes live, *over the top* (OTT) streaming for broadcast content, online gaming and personal live broadcast applications like

Facebook live. The workflow is quite similar to live video broadcasting in that real-time video encoding in fixed bandwidth networks is needed. However, the main difference from traditional broadcasting is that the client bitrate is not fixed but varies dynamically based on network conditions. Furthermore, different clients who are served the same video across different locations can have very different network conditions, hence channel bitrates. Thus, creating just one encoded version of the video cannot effectively serve all end users who have different and changing requirements. This problem is solved by using adaptive streaming where multiple versions of the same content are encoded at different bitrates and resolutions and streamed. A bitrate ladder is defined that has the same video encoded using different resolutions and bitrates. Usually, every version is CBR encoded. Different users are thus served one of these versions of the content, based on their network conditions. Moreover, every client's encoded version of the video can also be dynamically switched to a higher or lower bitrate during playback to suit the changing network conditions.

11.4.3   VIDEO ON DEMAND STREAMING

Video on demand (VOD) applications are becoming increasingly popular thanks to services like Netflix, Amazon, and Hulu. The encoded content is stored online and accessed by the user in real-time playback. VOD is similar to live internet video streaming as its video streamed over the internet and uses adaptive encoding. However, the encoding is done offline and multiple passes and non-real time encoding is preferred, in order to increase the video quality. Multi-pass encoding with capped VBR can be well suited for this application.

11.4.4   STORAGE

This is used by enterprise and private users to store encoded video in personal drives or cloud storage for archival purposes. The goal is to achieve the best possible quality without too much concern about the file size. Real-time or non-real time CRF encoding would be a good encoding mode to use for this class of applications. If, however, devices like DVDs or Blu Ray Disks are used for storage, there are fixed size restrictions that also have to be considered. The encoding in these cases is done with capped VBR mode, either in real time or non-real time, using some form of multi-pass encoding.

11.5   SUMMARY

- Different application scenarios define how encoders allocate bits to frames. Three important bitrate modes are: 1) CBR, 2) VBR, and 3) CRF.
- In CBR encoding, the encoder imposes more rigorous constraints on the bitrate around periodic intervals and encodes at a more or less consistent rate by disallowing dramatic bitrate swings.
- In VBR mode, the encoder allows more bits as necessary to the more complex segments of the video and uses fewer bits to encode simpler and more static segments.
- The CRF mode is a new mode. It is a constant quality encoding mode that prioritizes the quality metric and ensures a fixed quality across all sections of the video.
- CBR is a more predictable mode that is compatible across wider variety of systems compared to VBR mode.

12 PERFORMANCE

Video encoding is often an irreversible, lossy process wherein the encoded video is a good approximation of the source and the quality of this approximation depends on various encoding parameters like the quantization parameter (QP) that we discussed earlier in this book. Presumably, the encoded video is degraded relative to the source. The quality of encoding is gauged using a measure of this perceived video degradation compared to the source. The distortion or artifacts produced by the encoding process negatively impact the user experience and this is of paramount importance for content providers and service providers who deploy these systems.

The most important characteristic of any video encoder is quality. Any video encoder goes through evaluations to assess how it performs by using input video sequences that represent a broad variety of content and analyzing the encoded outputs. These clips are typically encoded using standard settings at various target bitrates. There are two broad ways to evaluate the output video quality:

1. **Objective Analysis**. This uses mathematical models that approximate a subjective quality assessment. Assessments are automatically calculated using a computer program. The advantage of using this method is that it is easily quantified and always provides a uniform and consistent result for a given set of outputs and inputs. However, its limitations are usually in terms of how accurately the model can approximate human perception. While there are several metrics for objective analysis, three tools that are increasingly used in the industry, namely, PSNR, SSIM and VMAF are discussed in this chapter.

2. **Subjective Analysis**. Here, the set of test video clips is shown to a group of viewers and their feedback, which is usually in some form of a scoring system, is averaged into a mean opinion score (MOS). While this method is not easily quantifiable, it's the most frequently used method.

   This is because it's simpler than objective analysis and it connects directly to the real-world experiences of users, who are the ultimate judge of perceived quality. However, the testing procedure may vary depending on what testing setup is available, what encoders are used for the testing, and so on. Subjective analysis is also prone to user bias and opinions.

12.1   OBJECTIVE VIDEO QUALITY METRIC

There has been an increasing need to develop objective quality measurement techniques because they provide video developers, standards organizations and other enterprises with the tools to evaluate video quality automatically without the need to view the video. In addition to the ability to benchmark video algorithms, the objective metrics can also be embedded into video coding pipelines as part of the algorithms to optimize and fine tune the quality during the encoding process itself. It should be noted that a majority of the objective *video quality* (VQ) metrics assume that the undistorted source is available for analysis. Such metrics are called *full reference* (FR) VQ metrics, the most common of which is the *peak signal-to-noise ratio* (PSNR) metric. This is widely used as it's simple to calculate and can be easily integrated within algorithms for optimization. However, PSNR has its limitations. It sometimes does not correlate well with perceived video quality, meaning that a video can look visually good but still have a poor PSNR value and vice versa. Despite such inherent limitations it remains one of the easiest and most widely used metrics in the industry. In this section, we will review three such objective metrics and consider the advantages and disadvantages of each.

12.1.1   PEAK SIGNAL-TO-NOISE RATIO (PSNR)

PSNR is the ratio between the maximum power of an input signal and the power of compression error (noise). It is expressed in a logarithmic scale. The metric usually provides a good approximation of the perceived quality of the reconstructed output. A higher PSNR value corresponds to higher visual quality. The denominator in the PSNR ratio involves the power of the compression error which is computed using the *mean squared error* (MSE) between the source (or reference) and the encoded (and subsequently reconstructed) video. PSNR is thus a simple function of this MSE value.

The PSNR is defined as:
$$PSNR = 10. \log_{10} (MAX^2 / MSE)$$
where MSE is defined as:

$$\text{MSE} = \frac{1}{m.n}\sum_i \sum_j |\text{Orig}_{i,j} - \text{Recon}_{i,j}|^2$$

The MSE is thus the average of the squared error between the original source pixels and the reconstructed pixels. MAX is the maximum pixel value, which is 255 for a bit depth of 8. The PSNR computation is usually applied frame-by-frame on all the components (especially luma) and the average value for the entire video sequence is used. Typical values for the PSNR in video encoding are between 25 and 50 dB for luma and this depends on the video content, bit rate and QP values used for encoding. Based on the HVS, a PSNR of 45dB and above usually corresponds to imperceivable visual quality impact in the video.

12.1.2   STRUCTURAL SIMILARITY (SSIM)

SSIM is a popular method for quality assessment of still images. It was first proposed for images [1] and has been extended to video. Like PSNR, SSIM is another *full reference* (FR) metric used for measuring the similarity between two images and is designed to improve on traditional methods such as PSNR and MSE. The idea behind SSIM is that the HVS is highly specialized in extracting *structural information* from visual content and the better the encoder preserves the structural information, the higher is the perceived visual quality. Traditional methods like PSNR focus instead on pixel residual errors. The HVS does not extract these from images; hence, they may not directly correlate to perceived quality. A metric based on structural distortion would have a high correlation with visual quality as perceived by the HVS. This would be a better option, in that it effectively blends subjective and objective testing.

Figure 90, from Wang, Lu, & Bovik [1], illustrates the above concepts. The original *Goldhill* image in Figure 90 (a) has been subjected to the following: (b) distortions including global contrast suppression, (c) drastic JPEG compression, and (d) blurring. In these tests, all the images were set up for a similar MSE relative to the source image.

This means that each of these images would yield similar PSNR values. It is visually obvious from the pictures that, despite the similarity in MSE, the distorted images are very different. A glance is sufficient to perceive that picture (b) is far more visually appealing than the other distorted images. In the JPEG compressed (c) and blurred (d) images, hardly any structures of the original image are preserved, hence they can't be seen, either. On the other hand, the image structures are preserved in the contrast-suppressed image (b). The point here is that error-based metrics like PSNR are prone to fail in scenarios like this and can be misleading.

Figure 90: Comparison of images with similar PSNRs but different structural content.

SSIM, on the other hand, does away with error-based computations. Instead, it leverages the characteristic of the HVS to focus on *structural information*. SSIM defines a model to measure image quality degradation based on changes in its structural information.

The idea of structural information is that the strong spatial correlations among pixels in an image or video picture carry important information about the structure of the objects in the picture. This is ignored by error-based metrics like PSNR that treat every pixel independently. If x = {xi | i = 1, 2,...,N} is the original signal and y = {yi | i = 1,2,...,N} is the reconstructed signal, then the SSIM index is calculated using the following formula[1]:

$$SSIM = \frac{(2x_m y_m + A)(2\sigma xy + B)}{(x_m^2 + y_m^2 + A)(\sigma_x^2 + \sigma_y^2 + B)}$$

In this equation, x_m and y_m are the mean of x and y, respectively, and σ_x, σ_y, σ_{xy} are the variance of x, the variance of y and the covariance of x and y. A and B are constants that are defined based on bit depths. The value of SSIM ranges from 0 to 1 with 1 being the best value. A 1 means that the reconstructed image is identical to the original image.

In general, SSIM scores of 0.95 and above are found to have imperceivable visual quality impact (similar to PSNR greater than 45dB). As with PSNR, the SSIM index is computed frame-by-frame on all three components of the video separately and the overall SSIM index for the video (for every component) is computed as the average of all the frame values.

12.1.3   VIDEO MULTIMETHOD ASSESSMENT FUSION (VMAF)

Video content is highly diversified and different kinds of distortions are possible across a variety of content. A video quality metric like PSNR while suitable for certain source and error characteristics, might not provide optimal assessment for others. An emerging trend, thus, is to combine the major existing metrics and use them in combination to derive a fusion method. Novel machine learning tools can be used to assign weights to the different elementary metrics in accordance with the source content and artifact characteristics. Using the existing objective methods, features from the video are extracted and used to feed machine learning algorithms to obtain a trained model that is used to predict the perceived VQ. While such a method does increase the complexity, it has been demonstrated to achieve significantly better performance. Furthermore, the performance of the learning tools can also be continuously improved.

Prof C.-C. Jay Kuo and his colleagues studied ten existing and better-recognized objective quality metrics and 17 image distortion types (such as JPEG compression distortion, quantization noise, and so on) [2]. It was observed that different quality metrics work well with respect to different image distortion types. For example, PSNR may not accurately measure quality for many distortion types, but it works well for additive noise and quantization noise distortions. In general, PSNR-based metrics were found to work well for half of the distortion types, while a feature similarity index metric worked well for the remaining distortions. Thus, they were motivated to develop a unified method that handles all the different distortions by fusing the distortion indices from the preferred elementary metrics into one final score.

Video multimethod assessment fusion (VMAF) is one such fusion metric that has been developed by Netflix in cooperation with Prof C.-C. Jay Kuo and his team at the University of Southern California. It has garnered the interest of the video community in recent times. VMAF computes an index that quantifies *subjective* quality by combining three elementary VQ metrics using a machine learning algorithm called *support vector machine* (SVM) regressor [3]. The SVM algorithm assigns weights dynamically to each elementary metric to derive the final metric. As it preserves and leverages the strengths of each metric, the SVM metric is considered a better estimate of perceived subjective quality. For example, if the correlation between the subjective *mean opinion score* (MOS) and an elementary metric is high, the SVM may assign a higher weight and vice versa. The machine-learning model can be trained using subjective experimental data such that the weights produced and the resulting VMAF index across a variety of content accurately reflect perceptual quality.

The Netflix VMAF algorithm [3] incorporated two image quality elementary metrics, namely, *visual information fidelity* (VIF) and *detail loss metric* (DLM). It also incorporated motion information by calculating the luma mean absolute pixel differences to account for the temporal characteristics. The above metrics are fused using the SVM regression algorithm to provide a single score for every frame. This is then averaged over all the videos to derive the final overall differential MOS (DMOS) value for the entire sequence. It's important to also develop a subjective testing method that yields MOS data that can be used by VMAF to train the internal machine learning model. Also, by using this framework, application-specific customized fusion VMAF metrics can be implemented by experimenting with other elementary metrics, features, and different machine learning algorithms. Some studies [4] have shown that a VMAF score of 93 and above results in imperceivable VQ impact and at lower scores, a change of about 6 VMAF points results in noticeable VQ impact. Furthermore, it is also understood that lower resolution videos have VMAF scores a lot worse than higher resolution videos.

12.2   ENCODER IMPLEMENTATIONS

There are a number of implementations available for the various codecs we have discussed. H.264 has naturally been the most adopted and widely used. In this section, we shall continue to focus on the three important codecs. These offer a rich insight in to the coding tools and implementation options available at the present time. Each of these three codecs has at least one freely downloadable source that can be compiled and run using the command line.

Alternatively, they can also be used under the widely popular FFmpeg framework if it has been compiled with the corresponding codec support. Using FFmpeg may offer much more flexibility and more options under a unified framework. This may be useful to some who are familiar with the tool and for whom having a standalone compiled codec executable is just fine for the purposes of exploring the various codec tool options.

In terms of codec comparisons, various results of comparing newer codecs like VP9 and H.265 to the existing and dominant H.264 are widely published and available on the internet. These have to be carefully analyzed as some of these comparisons may be outdated. Updated results with improved versions of the codecs may not be available. In general, it's hard to compare different codecs as they usually have different tool sets and not all tools are available across different implementations. This can lead to non-apples-to-apples comparison scenarios. When comparing codecs, the following three aspects are considered for a holistic picture.

1. Compression efficiency offered by the encoder across different settings
2. Encoding speed
3. Decoding speed

In this section, we shall focus exclusively on encoding compression efficiency tools. These in general are the primary motivation for new codec development and different encoder implementations. Each of the implementations discussed below are industry de facto references that are used to benchmark other implementations.

12.2.1 H.264 ENCODER

The widely used implementation for H.264 is the *x264* software library developed by VideoLAN. It is available as a free download and is released under the terms of the GNU GPL. Another popular implementation is the open h264 software developed by Cisco Systems. This is open sourced and available under the BSD license. The use of these software implementations for internet video is free as announced by MPEG LA, the private organization that administers the licenses for patents applying to the H.264 standard.

12.2.2   H.265 ENCODER

x265 is the most popular publicly available implementation of the H.265 standard. The x265 project builds on source code from x264 and the source code for this project was made publicly available by a company called MulticoreWare in 2013. x265 is offered under either GNU GPL or a commercial license, similar to x264. x265 has been integrated into popular APIs including FFmpeg and Handbrake among others and, like x264, it can be run as a standalone executable or using the FFmpeg APIs. The x265 implementation has been widely compared for compression efficiency to earlier x264 and also VP9 under various comparison studies at different times. It has consistently performed well under different test conditions and different metrics, including SSIM and the new VMAF metric.

12.2.3   VP9 ENCODER

VP9 source code (a.k.a. libvpx) is available under Google's WebM project. This can be downloaded, compiled and run in the command line using the executable *vpxenc*. The documentation for VP9 is provided in the WebM project Wiki that also provides few VP9-specific settings for VOD, DASH, constant quality, and constrained quality. The implementation provides encoding using VBR, CBR, constant quality and constrained quality and both 1-pass or 2-pass encoding modes are available. It should be noted that not all libvpx configuration modes offer real-time encoding. VP9 bitstreams encoded using libvpx are containerized in WebM format. This is a subset of the Matroska container format.

12.3   VIDEO QUALITY ASSESSMENTS

As mentioned earlier, there are several published testing results for compression efficiency of various codecs, especially H.264, H.265 and VP9. Some of these are starkly different from the others in terms of their actual results and conclusions. The test process usually involves encoding a select set of video clips of varying content using standardized settings across a range of bit rates and then measuring the objective quality using a metric like PSNR, SSIM or VMAF. This is done for all the test codecs and the results are plotted using bit rate-to-objective quality graphs and compared. As SSIM is often considered a much better balance between subjective and objective testing, let us use it for our tests.

In general, engineers and students seeking to run tests to evaluate encoders can use the following steps:

1.   Carefully select the set of test clips to represent a variety of content (high motion, detailed textures, static, movie, animation, noisy content, and so on) and resolutions (typically UHD, HD, SD and down to a few lower resolutions). For every clip at every resolution, the following steps need to be carried out.
2.   Decide the application configurations and parameter set that need to be tested (e.g., live, offline, GOP settings, and so on).
3.   Select the encoders/codecs to be tested (e.g., VP9 reference, H.264 reference, and so on).
4.   Determine the objective metric that needs to be measured (e.g., PSNR, SSIM, and so on). One or more metrics can be used.
5.   Determine the suitable QP or bit rate range to run the tests.
6.   Write the command line scripts for each of these encoder runs. Although an identical command line configuration might not be possible across different codecs, as they have different tool sets, care should be taken in this step to keep the tests as similar as possible.
7.   Run the tests, preferably using automated scripts.
8.   Get the objective metrics like SSIM and also the corresponding output file sizes. The metric values for just luma can be used or, if preferred, an average of all three components can also be used. Normalize the metrics using the output file sizes to derive the bitrate-adjusted metrics.
9.   Draw the charts and most importantly, observe the trends and variations to derive useful conclusions.

In this section we shall illustrate how this is done using one example. The objective and scope of tests here is not to compare the various codecs but to illustrate the testing procedures that can be used to derive useful conclusions.

We run two of the codecs outlined earlier, namely, x264 and x265. We use the settings that correspond to highest quality (very slow settings) at a few select bitrates in the range from 1 Mbps to 8 Mbps for the first 200 frames of the *Stockholm* [5] HD 720p clip. The clip was downloaded in y4m format and converted to yuv format using FFmpeg.

The following are the command lines used for the sample tests across all the bitrates.

```
x264 Configuration

./x264 720p5994_stockholm_ter.yuv --output
1000_720p5994_stockholm_ter.264 --input-res 1280x720 --seek 0 --frames 200 -
-input-depth 8 --ssim --preset veryslow --no-scenecut --tune ssim --keyint -1 --
min-keyint 60 --fps 59.94 --bitrate 1000

x265 Configuration

./x265 720p5994_stockholm_ter.yuv --output
1000_720p5994_stockholm_ter.265 --input-res 1280x720 --seek 0 --frames 200 -
-input-depth 8 --ssim --preset veryslow --no-scenecut --tune ssim --keyint -1 --
min-keyint 60 --fps 59.94 --bitrate 1000
```

Table 15 consolidates the SSIMs and bitrates for each of the test runs. This is graphically plotted in Figure 91. The figure provides a visual comparison of SSIM vs. bitrates for both x264 and x265 encoders.

Table 15: SSIM for x264 and x265 encoding.

Bitrate (kbps)	SSIM (dB)	
	x264	x265
1000	8.076	9.137
2000	9.436	9.974
3000	9.876	10.282
4500	10.223	10.514
6000	10.442	10.663
8000	10.648	10.811

We see that the x265 curve is to the left of the x264 curve. This means that x265 provides a higher value of SSIM, hence higher perceived visual quality at the same bitrate as x264. Also, for any given SSIM value the leftmost curve x265 has a lower corresponding bitrate than the x264 curve. It can be verified by drawing a horizontal line similar to the line across the 10 dB SSIM point in the chart. This is a measure of relative compression efficiency across different encoders.

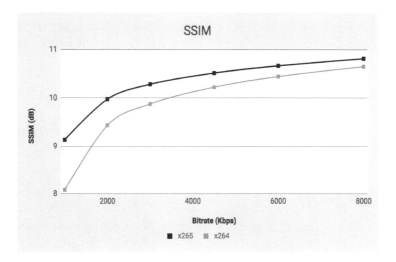

Figure 91: Comparison SSIM vs bit rates for x264 and x265 encoding.

12.4 SUMMARY

- There are two broad ways of evaluating output video quality: 1) objective analysis and 2) subjective analysis.
- A majority of the objective VQ metrics assume that the undistorted source is available for analysis and such metrics are called full reference (FR) metrics. These include PSNR, SSIM and VMAF.
- PSNR is the most widely used objective metric and is a simple function of the mean squared error (MSE) value between the source and the encoded video.
- The HVS is highly specialized in extracting structural information. While traditional methods like PSNR rely on extracting errors, SSIM focuses on extracting structural information.
- VMAF does not rely on a single method of distortion analysis. Instead, it uses existing metrics and combines them dynamically using machine learning tools.

- x264 and x265 are widely-used, downloadable software H.264 and H.265 encoders, respectively. vpxenc from the WebM project is the free-to-use VP9 encoder.
- Several quality comparisons exist that compare x264, x265 and libvpx. Both x265 and libvpx use newer tools and have performed well and significantly better than earlier-generation H.264 encoding.

12.5   NOTES

1.   Wang Z, Lu L, Bovik AC. Video quality assessment based on structural distortion measurement. *Signal Process Image Commun.* 2004;19(1):1-9. https://live.ece.utexas.edu/publications/2004/zwang_vssim_spim_2 004.pdf. Accessed September 22, 2018.

2.   Liu T, Lin W, Kuo C. Image quality assessment using multi-method fusion. *IEEE Trans Image Process.* 2013;22(5):1793-1807. https://www.researchgate.net/publication/234047751_Image_Qual ity_Assessment_Using_Multi-Method_Fusion. Accessed September 22, 2018.

3.   Li Z, Aaron A, Katsavounidis I, et al. Toward a practical perceptual video quality metric. The Netflix Tech Blog. https://medium.com/netflix-techblog/toward-a-practical-perceptual-video-quality-metric-653f208b9652. Published June 6, 2016. Accessed September 22, 2018.

4.   Rassool R. VMAF Reproducibility: Validating a Perceptual Practical Video Quality Metric. Real Networks. https://www.realnetworks.com/sites/default/files/vmaf_reproduci bility_ieee.pdf. Accessed October 22, 2018.

5.   stockholm. xiph.org. Xiph.org Video Test Media [derf's collection]. https://media.xiph.org/video/derf/. Accessed September 22, 2018.

13 ADVANCES IN VIDEO

According to reports available on the internet, [1] video data is poised to consume up to 82% of Internet traffic by 2021 with live and VOD video, surveillance video, and VR video content driving much of the video traffic over the web. As more users increase their online video consumption and with the advent of infrastructural changes like 5G, high-quality video experiences with UHD and higher resolutions, frame rates with ultralow latencies will soon be everyday realities. This will be possible with a combination of advances in a few key areas that, cumulatively, are well suited to drive significant growth. In this chapter we will focus on the following three broad areas:

- Advances in machine learning and optimization tools that are being integrated into existing video encoding frameworks to achieve compression gains using proven and deployed codecs.
- Newer compression codecs with enhanced tools to address upcoming video requirements like increased resolutions. We will highlight coding tools in an upcoming next-generation coding standard called AV1.
- Newer experiential platforms like VR and 360 Video and their inherent video requirements that are important topics of upcoming research.

13.1 PER-TITLE ENCODER OPTIMIZATION

Per-title encoding has been around conceptually and in experimental stages for several years. It was deployed at scale by Netflix in December 2015 as outlined in a Netflix tech blog article [2] that has also inspired this section. Internet streaming video services traditionally use a set of bit rate-resolution pairs, a.k.a a bit rate ladder. This is a table that specifies, for a given codec, what bit rates are sufficient to use for any fixed resolution. The ladder also, therefore, defines at what bitrate transitions from one resolution to the other occur. For example, if the bitrate ladder defines 1280x720p at 1 Mbps and 720x480p at 500 kbps, then as long as the bitrate remains around 1 Mbps and above, the streaming would use 720p encoded stream. When the network conditions drop the available bitrate to below 1 Mbps, the streaming would be using the 480p version.

This implementation is called a fixed ladder, as the resolution used for every bitrate is always fixed. While this is easy to implement, a fixed ladder may not always be the optimal approach. For example, if the content or scene is simple with less texture or motion, it will still be encoded at a fixed bitrate that may be higher than what it really needs.

Conversely, high complex content or scenes may need more bits than what's allocated using even the highest bitrate in the fixed ladder. Also, for a given bit rate a better resolution could be chosen based on the *complexity of content* instead of a fixed resolution ladder. For example, complex scenes can be better encoded using 1280x720p at 2 Mbps while easier content can be encoded using 1920x1080p at the same bitrate. Thus, the fixed approach, while providing good quality, cannot guarantee the best quality for any specific content at the requested bitrate. It is obvious from these examples that the key to successful encoding that's missing in traditional fixed bitrate-resolution ladders is *content complexity.*

The concept can be explained using the following example, where a single source is encoded at three different resolutions starting with lower and moving to higher resolutions across various bitrates.

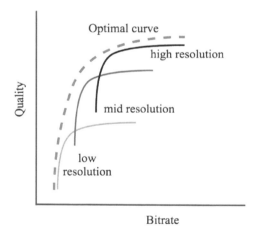

Figure 92: PSNR-bitrate optimal curve for encoding at three resolutions and various bitrates.

From Figure 92, adapted from the Netflix article, [2] we see that at each resolution, the quality gains start to diminish beyond a certain bitrate threshold. This means that beyond this point, the perceptual quality gains are negligible. This level, as seen in Figure 92, is different for different resolutions. It is clear from this chart that different resolutions are optimal at different bitrate ranges. This optimal bitrate is shown in the dotted curve that is the point of ideal operation. Selection of bitrate-resolution pairs close to this curve yields the best compression efficiency. Also, the charts are content specific and the optimal bitrates from this chart will not necessarily be optimal for other content. Per-title encoding overcomes this problem posed by fixed ladders by choosing content-specific bitrate-resolution pairs close to this curve for every title. To do this, experimental results from several data sets can be used to classify source material based on different complexity types and different bitrate ladders can be chosen per-title based on their content classification.

13.2   MACHINE LEARNING

Machine learning (ML) is the new-age technology that increasingly has an impact on everything, including video coding. ML tools have been around for several decades, since the time when researchers began studying ways to analyze and learn from data, using results of the analysis to build models and make predictions. Simply put, ML algorithms take a set of sample data as input, analyze the data samples, *learn* from them and make predictions based on what is learned to model systems. As more data gets input and processed, the output results are provided as feedback to the algorithms. This continuously trains the algorithms and improves the accuracy of the prediction models. This *learning* becomes a key aspect of the overall ML framework. In today's data age, the ML approach thus provides a mechanism to model any complex scenario solely by processing large volumes of data and learning from the observed outputs. Let us see in this section how and where ML tools can be used in video coding.

13.2.1   ML TOOLS FOR VIDEO CODING OPTIMIZATION

An ML system can be built and trained using a set of data. This system can then model and provide predicted outputs for future inputs. While a variety of predicted outputs are possible for any given set of inputs, these algorithms use mathematical optimization techniques that minimize a cost function to get the best output. Thus, ML tools can be deployed to provide solutions for any kind of optimization problem and this includes video coding. We established in Chapter 10 how the fundamental encoding problem is an optimization problem, one of deriving an optimized video quality for a given bit rate and vice versa.

Video optimizations using existing codecs like H.264 and VP9 are becoming more important as more and more streaming service providers are using optimization as a key tool to differentiate themselves from competitors and engage their users by improving the user experience. Video coding optimization-driven bit rate savings also mean little change to existing infrastructure and workflows. These are huge benefits to any video service provider.

Early in 2018, Netflix announced their dynamic optimizer implementation [3]. Their algorithm used machine learning techniques to analyze each frame and used compression based on the content. The VMAF quality metric was used as the optimization objective and every shot was encoded at different bitrates to meet the target VMAF objective. In this approach, a variety of shots were shown to users who rated them for complexity and content. The resulting subjective scores from these tests (MOS) were then used to train ML algorithms to model the picture quality. This was in turn used to optimize encoding of the video sequence on a shot-by-shot basis. It was observed that using this approach, the overall bit rate was considerably reduced while still maintaining a uniform quality level, as every shot was optimally encoded. Netflix was able to demonstrate streams that looked identical at half the bandwidth using this optimization technique, especially at low bitrates.

Other ideas to reduce encoding complexity using ML tools have also been explored by G. F. Escribano and his colleagues [4]. They employed low-complexity input video attributes to train an ML algorithm that can then classify video coding decisions and obtain a video coding decision tree using features derived from input video.

Common attributes used in video coding algorithms include statistical pixel mean, variance values, SAD values and the like. The computationally simple coding decision trees can then effectively supplement or, in some cases, even replace computationally expensive cost-based decisions. The fundamental challenges here would be around the selection of the low complexity input video attributes and also the selection of the classification ML algorithms. The process is illustrated in Figure 93.

Needless to add, the accuracy of the ML algorithm outputs would be highly dependent on the selection of the input data set that is used for training purposes.

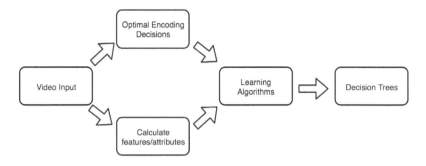

Figure 93: Applying machine learning to build mode decision trees.

13.3   EMERGING AV1 CODEC

AV1 is an emerging, open and royalty-free standard developed by the Alliance for Open Media (AOM), a consortium of several companies in the video industry. The AV1 format is based on Google's VP10 project and is expected to produce anywhere from 25-40% efficiency improvements over existing codecs like VP9 and H.265. It is specifically designed for web streaming applications with higher resolutions and frame rates and HDR functions. The reference encoder is available for free download online and the work on improving the computational efficiency to make encoding runs faster is ongoing. AV1 uses the same traditional, block-based, hybrid model with several enhancements that, in aggregate, account for its increased compression efficiency. In this section, we will explore the enhancements and features offered by AV1. Table 16 best illustrates these enhancements and how they compare to industry-leading HEVC and VP9 standards.

Table 16: Comparison of AV1 tools and enhancements against HEVC and VP9.

Coding Tool	VP9	HEVC	AV1
Block size	64x64	Variable 8x8 to	128x128

Coding Tool	VP9	HEVC	AV1
	superblocks	64x64 CTUs	superblocks
Partitioning	Variable partitions from 64x64 hierarchically down to 4x4	Variable partitions from 64x64 hierarchically down to 4x4	Additional *T-shaped* partitioning schemes supported. Also, wedge-shaped partitioning for prediction is introduced to enable more accurate separation of objects along non-square lines.
Transforms	Variable 32x32 down to 4x4 integer DCT transforms + 4x4 integer DST transform	Variable 32x32 down to 4x4 integer DCT transforms + 4x4 integer DST transform	Variable 32x32 down to 4x4 transforms. Rectangular versions of the DCT and asymmetric versions of the DST. combine two 1-D transforms to use different H & V transforms.
Intra prediction	10 modes including 8 angles for directional	35 direction for prediction	65 angles for prediction and includes weighted

Coding Tool	VP9	HEVC	AV1
	prediction		prediction.
Inter prediction	No weighted prediction support	Supports temporal weighted prediction.	Supports spatial overlapped block motion compensation, global MVs, and warped MVs.
Sub pixel interpolation	Support for higher precision ⅛ pel MVs	¼ pixel eight-tap filter Y and ⅛ pixel four-tap UV	Support for higher precision ⅛ pel MVs
Internal precision for prediction	Configurable 8-bit and 10-bit are supported	Configurable 8-bit and 10-bit are supported	Internal processing in higher precision configurable as 10 or 12 bits
Filtering	In-loop deblocking filter affecting up to 7 pixels on either side of the edges. No SAO filter.	In-loop deblocking filter affecting up to 3 pixels on either side of the edges. SAO filter with sample and band offset modes.	Loop restoration filter similar to deblocking filter. Also includes a new directional filter to remove ringing noise.
Entropy coding	Binary arithmetic coding with frame level adaptation	Binary arithmetic coding with row level adaptation	Non-binary arithmetic coding with symbol level probability adaptation.

Coding Tool	VP9	HEVC	AV1
Block skip modes	No support for skip modes	Merge modes	No explicit skip modes
Motion vector prediction	4 modes including three modes with implicit MV prediction from spatial and temporal neighbors.	Enhanced spatial and temporal prediction	Dynamic reference MV selection [5]
Parallelism tools	No WPP but column and row tiles are supported with no prediction across column tiles and prediction possible across tile rows.	Wavefront parallel processing, tiles, slices	No WPP but column and row tiles are supported with configurable prediction across tile rows.
Reference pictures	Up to 3 frames from 8 available buffers	Up to 16 frames depending on resolutions	Up to 6 frames from 8 available buffers
Interlaced coding	Only frame coding is supported.	Only frame coding is supported.	Only frame coding is supported.

13.4   VIRTUAL REALITY AND 360° VIDEO

Emerging technologies like 360° video (immersive or spherical video) and VR aim to provide a digital experience in a fully immersive environment with full fidelity to human perception. In a regular video or movie experience, cameras capturing roughly 8 million pixels per frame (UHD) for each component at 30fps is perhaps perfectly adequate for a good experience.

However, for a completely immersive experience, the captures have to match the rates of capture by human eyes that receive light constantly and perceive much faster motion, say, between 90-150fps. Also, as the field of view at such close proximity is higher, a far greater number of pixels will need to be processed in every frame. By one estimate, the human eye can receive 720 million pixels for each of 2 eyes. [6] At 36 bits per pixel and 60fps, this adds up to 3.1 trillion terabits.

Even when compressed 600 times by existing compression schemes like H.265, this would require 5.2 Gbps. While the calculations may or may not be completely accurate, what is important here is to understand that fully immersive experiences are a completely different ball game compared to traditional video experiences.

Furthermore, unlike in traditional video where the image is projected on a flat plane, a 360 video is projected onto a spherical plane around the viewer. This presents challenges when such projected images are compressed using traditional coding tools that are built around conventional video.

13.4.1   WHAT IS 360° VIDEO?

In 360° videos, every direction or angle that's needed to complete the full 360° is simultaneously captured, either using only one single omnidirectional camera or a special rig of several cameras. The angles that need to be captured range horizontally from -180° to +180° and vertically from -90° to +90°. The captures from all these angles are then stitched to produce one final video that gives a complete spherical view of the scene.

This can be done either directly within the camera itself or externally using merging software. It is then projected either monoscopically or stereoscopically. The latter means different images, one for each eye.

The result can be viewed using a flat panel or a *head-mounted display* (HMD). Only a small section of the projection, called the viewport, is viewed at any time and the section being presented varies dynamically based on the head or device motion.

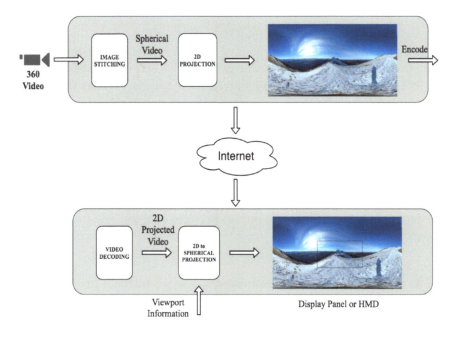

Figure 94: Current 360° video delivery workflow with 2D video encoding.

Image source: https://pixabay.com/en/winter-panorama-mountains-snow-2383930/

This workflow is illustrated in Figure 94, where current generation codecs like HEVC are used to encode the 2D video that is obtained after 2D projection of the input spherical video.

The projection information for standard projection formats is also available to be included in HEVC SEI messages. When the streaming video is received, it is decoded using standard HEVC decoders that also propagate the projection information. This projection information, along with the viewport information, is used to extract the specific section and display it spherically on the HMD.

The selection of the projection technique is extremely important as it determines how we finally perceive the three-dimensional space (spherical) view using a two-dimensional surface (plane) and there are different ways in which this can be done. Popular projection schemes include *equirectangular*, *cube map* and *pyramid* projections.

Amongst these, cube map and pyramid projection schemes have been found to provide significant data reduction compared to equirectangular projections.

Equirectangular is the simplest projection. Every longitudinal meridian is mapped to vertical straight lines with fixed spacing and every latitudinal circle is mapped to a horizontal straight line with fixed spacing. This introduces significant distortions, especially at the poles. Furthermore, it also creates warped images with bent motion. This is undesirable for block-based compression schemes that assume straight line motion vectors. Also, the top and bottom poles of the equirectangular projection contain redundant data that are expensive to process.

The problems with equirectangular projections can be addressed using cube maps. These have six unique faces in which pixels are uniformly distributed without any bending distortions within the faces themselves. This also means that there are no poles containing any redundant information.

While cube maps seem perfectly fine to solve problems with equirectangular projections, they use equal distribution across all angles (represented by the faces of the cube) and still consume a lot of bandwidth, especially for a VR experience using 360° where headsets can support resolutions beyond 4K. Encoding and streaming this whole set of data thus produces two problems.

1) There is enormous bandwidth, typically a few hundred Mbps.
2) Not all decoders can handle resolutions beyond UHD, especially mobile devices that are used on mobile VR headsets.

While it's true that devices offer only a limited field of view (FOV) of ~95° and much of the 360° is redundant, in VR the user controls the immersive experience and every slight change in the head and eye movement results in a different view that needs to be presented with utmost fidelity. Thus, the challenge here is to somehow transmit the complete FOV yet optimize the significant redundancies that exist in views that are outside the immediate FOV. Thus, 360° video poses significant challenges in processing the huge amounts of data. As compression schemes still rely on traditional block-based approaches that are built around conventional video, there are significant research opportunities to model the complete

360° field of view and associated motion to develop newer prediction modes that are suited for these upcoming applications and ecosystems.

The Joint Video Experts Team (JVET) is a joint effort between ISO-IEC and ITU-T. The team is now working on the successor to the HEVC codec called Versatile Video Coding or VVC. It targets the efficient encoding of 360° video content with known or new projection formats. This video standard is targeted to be finalized by 2020 and also aims to achieve 50% bit rate efficiency over HEVC for traditional 2D content. Some 360° video-specific tools that have been proposed as part of this standard include prediction using reference samples based on spherical geometry instead of the regular 2D rectangular geometry, adaptive quantization based on spherical pixel density, and others. The compression benefits would result in an immediate reduction in storage costs and bandwidth reduction that will be needed for at-scale deployments.

13.5 SUMMARY

- Per-title encoding overcomes the limitations of a fixed resolution encoding ladder by incorporating content complexity as a measure to determine at what bitrate and resolution a specific content will be encoded.
- At the core of ML prediction algorithms are mathematical optimization techniques that minimize an internal cost function to get the best possible predicted output among all others. This can be explored to optimize video encoding.
- AV1 is the new open, royalty-free standard developed by the Alliance for Open Media (AOM) and uses the block-based hybrid model by building on VP9 with several enhancements. These, in aggregate, account for its increased compression efficiency.
- Emerging technologies like 360 Video and Virtual Reality have more complex visual scenarios that require several magnitudes of higher video throughputs with low latency. These together put forth significant compression requirements and thus opportunities exist to develop compression systems suited for these applications.

13.6 NOTES

1. *Cisco Visual Networking Index: Forecast and Methodology, 2016–2021.*
    Cisco. June 6, 2017.
    https://www.cisco.com/c/en/us/solutions/collateral/service-
    provider/visual-networking-index-vni/complete-white-paper-c11-
    481360.html. Updated September 15, 2017. Accessed September 22,
    2018.
2. Aaron A, Li Z, Manohara M, et al. Per-title encode optimization. The
    Netflix Tech Blog. https://medium.com/netflix-techblog/per-title-
    encode-optimization-7e99442b62a2. Published December 14, 2015.
    Accessed September 22, 2018.
3. Jones A. Netflix introduces dynamic optimizer. Bizety.
    https://www.bizety.com/2018/03/15/netflix-introduces-dynamic-
    optimizer/. Published March 15, 2018. Accessed September 22,
    2018.
4. Escribano GF, Jillani RM, Holder C, Cuenca P. Video encoding and
    transcoding using machine learning. *MDM'08 Proceedings of the 9th
    International Workshop Multimedia Data Mining: Held in Conjunction
    with the ACM SIGKDD 2008*. New York, NY: Conference KDD'08 ACM;
    2008:53-62.
    https://www.researchgate.net/publication/234810713/download.
    Accessed September 22, 2018.
5. Massimino P. AOM - AV1 How does it work? AOM-AV1 Video Tech
    Meetup. https://parisvideotech.com/wp-
    content/uploads/2017/07/AOM-AV1-Video-Tech-meet-up.pdf.
    Published July, 2017. Accessed September 22, 2018.
6. Begole, B. Why the internet pipes will burst when virtual reality
    takes off. *Forbes Valley Voices.* Forbes Media LLC.
    https://www.forbes.com/sites/valleyvoices/2016/02/09/why-the-
    internet-pipes-will-burst-if-virtual-reality-takes-off/#34c7f6e43858.
    Published February 9, 2016. Accessed September 22, 2018.

RESOURCES

Bankoski J, Wilkins P, Xu X. Technical overview of VP8, an open source video codec for the web. Google, Inc. http://static.googleusercontent.com/media/research.google.com/en/us/pubs/archive/37073.pdf. Accessed September 23, 2018.

Bultje RS. Overview of the VP9 video codec. Random Thoughts Blog, General [Internet]. https://blogs.gnome.org/rbultje/2016/12/13/overview-of-the-vp9-video-codec/. Published December 13, 2016. Accessed September 23, 2018.

Exponential-Golomb Coding. Wikipedia. https://wikivisually.com/wiki/Exponential-Golomb_coding. Updated July 9, 2018. Accessed September 23, 2018.

Ghanbari M. History of video coding. Chapter 1 in *Standard Codecs: Image Compression to Advanced Video Coding*. London, England: Institution of Electrical Engineers; 2003. https://flylib.com/books/en/2.537.1/history_of_video_coding.html. Accessed September 23, 2018.

Grange A, Alvestrand HT. A VP9 bitstream overview. Network Working Group Internet-Draft. ITEF.org. https://tools.ietf.org/id/draft-grange-vp9-bitstream-00.html#rfc.section.2.6. Published February 18, 2013. Accessed September 23, 2018.

Grange A, de Rivaz P, Hunt J. VP9 bitstream and decoding process specification - v0.6. Google, Inc. webmproject.org. https://storage.googleapis.com/downloads.webmproject.org/docs/vp9/vp9-bitstream-specification-v0.6-20160331-draft.pdf. Published March 31, 2016. Accessed September 23, 2108.

Karwowski D, Grajek T, Klimaszewski K, et al. 20 years of progress in video compression – from MPEG-1 to MPEG-H HEVC. General view on the path of video coding development. In Choras RS, ed. *Image Processing and Communications Challenges 8: 8th International Conference, IP&C 2016 Bydgoszcz, Poland, September 2016 Proceedings*. Cham, Switzerland: Springer International Publishing; 2017:3-15.

https://www.researchgate.net/publication/310494503_20_Years_of_Pro
gress_in_Video_Compression_-_from_MPEG-1_to_MPEG-
H_HEVC_General_View_on_the_Path_of_Video_Coding_Development.
Accessed September 23, 2018.

Marpe D, Schwarz H, Wiegand T. Context-Based Adaptive Binary
Arithmetic Coding in the H.264/AVC Video Compression Standard. *IEEE
Circuits Syst Video Tech.*
http://iphome.hhi.de/wiegand/assets/pdfs/csvt_cabac_0305.pdf.
Accessed September 23, 2018.

Melanson M. Video coding concepts: Quantization. Breaking Eggs and
Making Omelettes: Topics on Multimedia Technology and Reverse
Engineering; Multimedia Mike [Internet].
https://multimedia.cx/eggs/video-coding-concepts-quantization/.
Published April 5, 2005. Accessed September 23, 2018.

Mukherjee D, Bankoski J, Bultje RS, et al. A technical overview of VP9—
the latest open-source video codec. *SMPTE Motion Imaging J.*
2015;124(1):44-54.
https://www.researchgate.net/publication/272399193/download.
Accessed September 22, 2018.

Mukherjee D, Bankoski J, Bultje RS, et al. A technical overview of VP9: The
latest royalty-free video codec from Google. Google, Inc.
http://files.meetup.com/9842252/Overview-VP9.pdf. Accessed
September 23, 2018.

Ozer J. Finding the just noticeable difference with Netflix VMAF.
Streaming Learning Center. streaminglearningcenter.com.
https://streaminglearningcenter.com/learning/mapping-ssim-vmaf-
scores-subjective-ratings.html. Published September 4, 2017. Accessed
September 23, 2018.

Ozer J. Mapping SSIM and VMAF scores to subjective ratings. Streaming
Learning Center. streaminglearningcenter.com.
https://streaminglearningcenter.com/learning/mapping-ssim-vmaf-
scores-subjective-ratings.html. Published July 5, 2018. Accessed
September 23, 2018.

Ozer J. *Video Encoding By The Numbers: Eliminate the Guesswork from Your Streaming Video*. Galax, VA: Doceo Publishing, Inc.; 2017.

pieter3d. How VP9 works, technical details & diagrams. Doom9's Forum [Internet]. https://forum.doom9.org/showthread.php?t=168947. Published October 8, 2013. Accessed September 23, 2018.

Rate distortion optimization for encoder control. Fraunhofer Institute for Telecommunications, Heinrich Hertz Institute, HHI. https://www.hhi.fraunhofer.de/en/departments/vca/research-groups/image-video-coding/research-topics/rate-distortion-optimization-rdo-for-encoder-control.html. Accessed September 23, 2018.

Riabtsev S. Video Compression. www.ramugedia.com. http://www.ramugedia.com/video-compression. Accessed September 23, 2018.

Richardson I. A short history of video coding. Invited talk at United States Patent and Trade Office, PETTP 2014 USPTO Tech Week, December 1-5, 2014. SlideShare Technology. slideshare.net. https://www.slideshare.net/vcodex/a-short-history-of-video-coding. Accessed September 23, 2018.

Richardson I, Bhat A. Historical timeline of video coding standards and formats. Vcodex. https://goo.gl/bqyyXd. Accessed September 23, 2018.

Richardson IE. *H.264 and MPEG-4 Video Compression: Video Coding for Next-generation Multimedia*. West Sussex, UK: John Wiley & Sons, Ltd.; 2003.

Robitza W. Understanding rate control modes (x264, x265, vpx). SLHCK.info. https://slhck.info/video/2017/03/01/rate-control.html. Published March 1, 2017. Updated August, 2018. Accessed September 23, 2018.

Sonnati F. Artificial Intelligence in video encoding optimization. Video Encoding & Streaming Technologies, Fabio Sonnati on Video Delivery and Encoding Blog [Internet]. https://sonnati.wordpress.com/2017/10/09/artificial-intelligence-in-video-encoding-optimization/. Published October 9, 2017. Accessed September 23, 2018.

Sullivan GJ, Ohm J, Han W, Wiegand T. Overview of the High Efficiency Video Coding (HEVC) Standard. *IEEE Trans Circuits Syst Video Technol.* 2012;22(12):1649-1668. https://ieeexplore.ieee.org/document/6316136/?part=1. Accessed September 21, 2018.

Urban J. Understanding video compression artifacts. biamp.com. Component. Biamp's Blog [Internet]. http://blog.biamp.com/understanding-video-compression-artifacts/. Published February 16, 2017. Accessed September 23, 2018.

Vinayagam M. Next generation broadcasting technology - video codec. SlideShare Technology. slideshare.net. https://www.slideshare.net/VinayagamMariappan1/video-codecs-62801463. Published June 7, 2016. Accessed September 23, 2018.

WebM Blog. webmproject.org. http://blog.webmproject.org. Accessed September 23, 2018.

Wiegand T, Sullivan GJ, Bjøntegaard G, Luthra A. Overview of the H.264/AVC Video Coding Standard. *IEEE Trans Circuits Syst Video Technol.* 2003;13(7):560-576. http://ip.hhi.de/imagecom_G1/assets/pdfs/csvt_overview_0305.pdf. Accessed September 21, 2018.

Wien M. *High Efficiency Video Coding: Coding Tools and Specification.* Berlin and Heidelberg, Germany: Springer-Verlag; 2015.

xiph.org. [YUV video sources]. Xiph.org Video Test Media [derf's collection]. https://media.xiph.org/video/derf/. Accessed September 23, 2018.

Ye Y. Recent trends and challenges in 360-degree video compression. Keynote presentation at IEEE International Conference on Multimedia and Expo (ICME), 9th Hot 3D Workshop. InterDigital Inc. SlideShare Technology. slideshare.net. https://www.slideshare.net/YanYe5/recent-trends-and-challenges-in-360degree-video-compression. Published August 1, 2018. Accessed September 23, 2018.

Zhang H, Au OC, Shi Y, et al. Improved sample adaptive offset for HEVC. *Proceedings of the 2013 Asia-Pacific Signal and Information Processing Association Annual Summit and Conference.* IEEE. http://www.apsipa.org/proceedings_2013/papers/142_PID2936291.pdf . Published 2013. Accessed September 23, 2018.

INDEX

ABOUT THE AUTHOR

Avinash Ramachandran has been working in video compression for over 15 years as a serial startup technologist, innovator, and speaker. His work on video coding has contributed to several patents in motion estimation, motion compensation and bitrate control algorithms. He is currently developing next-generation algorithms and products with H.265, VP9 and AV1 codecs at NGCodec Inc. A senior member of IEEE, he completed his Masters in Digital Signal Processing from the Indian Institute of Technology Madras in India and holds an MBA from the Richard Ivey School of Business in Canada.

www.ingramcontent.com/pod-product-compliance
Lightning Source LLC
Chambersburg PA
CBHW041151050326
40690CB00001B/431